The Molecular Biology of Gaia

The Molecular Biology of Gaia

George Ronald Williams

Columbia University Press
New York

Columbia University Press
Publishers Since 1893
New York   Chichester, West Sussex

Copyright © 1996 Columbia University Press
All rights reserved

Library of Congress Cataloging-in-Publication Data
Williams, George Ronald.
The molecular biology of Gaia / George Ronald Williams.
p. cm.
Includes bibliographical references (p. 183) and index.
ISBN 0–231–10512–6 (cloth). — ISBN 0–231–10513–4 (paper)
1. Gaia hypothesis. 2. Molecular biology. I. Title.
QH331.W495   1996
96–17485
CIP

Casebound editions of Columbia University Press
books are printed on permanent and durable acid-free paper.

Printed in the United States of America
c 10 9 8 7 6 5 4 3 2 1
p 10 9 8 7 6 5 4 3 2 1

Contents

Acknowledgments	vii
Prologue	ix

1. **The Question of Habitability** — 1

   - Extinctions — 3
   - Geochemical Shifts — 7
   - Ice Ages — 9
   - Biogeochemical Cycles — 14
   - Assimilation in the Carbon Cycle — 20
   - Regenerating the Nutrient Pool — 25

2. **Dimensions of the Anthropogenic Perturbation** — 36

   - Human Impact on the Carbon Cycle — 39
     (1) Atmospheric Carbon Dioxide; (2) Other Carbon Compounds of the Atmosphere
   - Human Impact on the Nitrogen Cycle — 52
   - Human Impact on the Phosphorus Cycle — 55
   - Human Impact on the Sulfur Cycle — 57
   - Acid Rain — 58
   - Ozone — 60
   - The Greenhouse Effect — 63

vi   *Contents*

3. The Life Boundary and Environmental Homeostasis   65

   The Basis for Stability   69
   (1) Luck; (2) Size; (3) Equilibrium; (4) Feedback
   Biochemical Processes in the Geochemical
      Feedback Loops   79
   (1) Oxygen; (2) Carbon Dioxide; (3) Sulfate and Clouds

4. Global Metabolism, Geophysiology, and
   Gaian Metaphors   103

   Global Metabolism: Description, Simile, or Metaphor?   104
   Gaia and Geophysiology   108
   Justifying the Gaian Metaphors   111
   Patterns of Metabolic Control   114
   (1) Feedback Inhibition; (2) Allosteric Modulation;
   (3) Enzymic Cascades; (4) Gene Regulation
   Organismic Homeostasis   120

5. Teleonomy and the Biological Critique of Gaia   123

   Habitability and Adaptation   123
   Adaptation in Molecular Terms   127
   The Evolution of Physiological Regulation   130
   Gaia, Adaptation, and Selection   134
   The Daisyworld Answer and Its Problems   138
   How Can a Daisyworld Originate?   142

6. Molecular Regulation and Global Metabolism   146

   Can Global Stability Be Linked to
      Physiological Homeostasis?   146
   Liebig's Law and Redfield's Ratios: Coordination at
      the Ecological Level   149
   The Molecular Basis of the Global Nitrogen Cycle   154
   The Molecular Control of Glutamine Synthetase   157
   The Molecular Control of Nitrate Reductase   167
   The Molecular Control of Nitrogenase   171
   Toward a Molecular Biology of Gaia   176

References   183
Index   205

Acknowledgments

David Schwartzman and Tyler Volk read this book in draft form and I am grateful to them for their helpful suggestions. The editorial staff of Columbia University Press were unfailing in their assistance and encouragement. My particular thanks are due to Connie Barlow, who was able to help both with the content and with the expression of my arguments.

This book is about the stability and habitability of Earth. Joyce Williams has made one corner of the planet stable and habitable for many years and my debt to her is incalculable.

Prologue

It is more than twenty years since James Lovelock first put forward the idea of treating the ecology of planet Earth as the physiology of a living entity, to which he gave the name of "Gaia." The idea has been widely influential among environmental activists, and Lovelock's lucid and persuasive expositions of the concept have had an extensive readership among the general public. On the other hand, the idea has had a much harder time establishing itself among other scientists.

That this would be the case among geochemists and oceanographers is not surprising. Their consciousness of the magnitude of the grand cycles of the inorganic planet inevitably biases them against any scheme that offers the biosphere any role other than that of a catalyst of processes that are determined by Earth's physics and chemistry. It is perhaps more surprising that the idea has had an equally chilly reception from biologists. The reasons for this coolness (or even hostility) are complex; in chapters 4 and 5 I examine them in some detail. Above all, the problem seems to arise because no serious attempt has been made to integrate the Gaian idea into the mainstream of contemporary biological thought. For some of Gaia's most articulate proponents such an attempt would be fruitless because they perceive that mainstream as incapable of incorporating the Gaian idea and therefore it is biology that must be changed. Perhaps so, but it is easier to speak of paradigm shifts than to effect them—particularly at a time when the current paradigm, far from being in trouble, is enjoying extraordinary success. Such is the case for biology which, in the fifty years since the elucidation of the structure of DNA by Watson and Crick, has developed ever more powerful explanatory schemes to account for more and more facets of organismic processes—first of metabolism, then of genetics, and now of development and behavior.

This book is not a "who-done-it" that starts with the finding of a dead body; it is a "what-does-it" that starts with the finding of life on a planet, Earth. I start where Lovelock started, asking about the planetary conditions that permit habitability. Again, I follow Lovelock in recognizing stability as a necessary condition for the persistence of life. A claim for stability of the global ecosystem does not mean very much if that system has never been stressed. In the first chapter I review the evidence—extinctions, geochemical shifts, and ice ages—that suggests that the global environment has undergone major perturbations. Indeed, Earth's habitability has changed, but, at least so far, the planet has remained continuously hospitable since life arose 3.8 billion years ago. This hospitality betokens a close fit between the resources of the geophysical environment and the requirements of the global biota. This fit is not static. The environment and the biota are linked by the chemical fluxes that are traditionally known as the biogeochemical cycles. In the language of the Gaia hypothesis these fluxes might be referred to as "global metabolism." The second half of chapter 1 is given over to a description of some of the more important of these cycles.

Heinrich Holland (1984) ends his review of the geological history of the atmosphere and oceans with the warning, "It remains to be seen whether this long record [of the continuity of life] will be sustained in the presence of modern man." It is the threat of novel human pressures that provides the extrascientific impetus to investigate global stability and habitability. This ethical imperative is the reason why, in chapter 2, I briefly turn aside from my quest for the ground of that stability in order to remind the reader of the quantitative aspects of the human threat to the stable operation of the global biogeochemical cycles, the human impact on global metabolism.

In chapter 3 I return to the quest for a biochemical understanding of the stability of the global ecosystem. I use Peter Weyl's categorization of the bases for stability (luck, size, chemical equilibria, and negative feedback). It is the negative feedback loops where the biota might be expected to play a part. I use the examples of three components of the atmosphere (oxygen, carbon dioxide, and dimethyl sulfide) that are strongly implicated in global habitability. In each case I find reason to believe that molecular details of biochemical processes are implicated in determining the concentration of these gases in Earth's atmosphere.

Up to that point in my attempt to determine the relative roles of biochemistry and geochemistry in determining the character of the global environment, I have alluded to the concept of Gaia only sparingly. In chapters 4 and 5 the question of the appropriateness of Gaian metaphors comes to center stage. My analysis of the controversy, in chapter 4, hinges on the distinction between geochemical feedback (governed by simple mass action principles) and biochemical metabolic control, in which flow rates are controlled by highly specific modulation of the activity of the enzymes catalyzing the flow. This specificity has its molecular basis in the specificity of interaction of substrate ligands with the elaborately folded polypeptide chains of the enzymes. In chapter 5 I recognize the genetic origin of these elaborate three-dimensional molecular structures as the root of the deep-seated objections to Gaia felt by many biologists. These complex structures have been selected in the long process of Darwinian evolution. And it is Darwinian selection that accounts for the appropriateness of these structures to their biological function. The argument may be summarized: no selection, no Gaia.

This strong argument clearly identifies (if I may mix my mythological metaphors) the Achilles heel of Gaia. It may well be fatal to the characterization of Gaia as an organism. But it does not deal with the strength of Gaia as metaphor, with the implicit recognition of the role of the biota in stabilizing the life-sustaining character of the global environment.

In chapter 6 I seek a way out of this quandary. Can one try a different approach—an approach that builds on the strengths of the two camps? Is it possible to take Lovelock's idea of geophysiology seriously and go on to suggest that, just as the physiology of cells and organisms is now understood in light of the underlying biochemistry, so the workings of the planetary ecosystem will be understood in terms of the molecular details of the relevant biological processes?

It is not clear that this aim can be achieved. Gaia may be a bad idea, or contemporary biology may be wrong headed; in either case no reconciliation will be found. And it would certainly be immodest of me to suggest that this work resolves the dilemma. What I have tried to do is raise the possibility that the Gaian idea is, at least in principle, capable of being viewed in terms with which most biologists, even those whose practice is strictly reductionist, can feel reasonably comfortable.

Many years of research will be needed to see whether the possibility can be instantiated.

In chapter 6 I suggest where to begin looking: among the molecular processes that introduce the simple molecules of the inorganic environment into the complex chemistry of living systems. These processes in the biogeochemical cycles I call *mobilization* and *assimilation*. Attention to these pathways of material flow has guided my search. It must be conceded, however, that to establish the role of a particular microprocess in the macrosystem (that is, to find a causal path from a change in the interatomic relationships in a particular enzyme to the level of cellular function) has never been easy. To extend that chain of cause and effect to the ecological level will be that much harder—and even more difficult when we move to the global level. But until some such instance of the role of molecular regulation in ecosystem behavior can be convincingly displayed, Gaia may remain a fringe idea in a biological counterculture.

And that would be a pity. In my encounters with senior undergraduates and with graduate students I have met much interest in Gaia but genuine puzzlement about how the idea could be tackled observationally or experimentally. Students seem instinctively to sense the problems of confirmation or falsification. This book is intended, in part, to encourage those just beginning to think about a life in science to take a new look at Gaia. Here I shall suggest that the disassembly approach that is so familiar to scientists (take it apart and see how it works) is not contraindicated in studies of planetary ecology. Rather, there is much to be done in more-or-less traditional ways, in the field and at the bench, to flesh out the Gaian metaphor. To do this will, inevitably, become an occupation for specialists. But such specialists will need to know how their findings fit into a bigger picture—what questions they may answer at far vaster scales.

A book such as this can be written only to point to a direction. No one can possess the kind of expertise in realms as far apart as oceanography and protein structure that would be required for an attempt at in-depth integration. We specialists, therefore, must learn to talk to one another, and I shall be well rewarded if I do no more than stimulate such conversation. I can only hope that such experts will not be too harsh in their judgment of the solecisms that, I fear, must inevitably find their way into any attempt at a multidisciplinary monograph (itself, perhaps, an oxymoron). As for mistakes, so for omissions. The bibliography is not intended to be exhaustive. I have supplemented my use of original sources with extensive use of authoritative reviews, but this does carry risks. I apologize to those whose favorite references have been left out. My aim, rather than either intensity or completeness, has been to write of, for instance, enzymology in a way that would interest an atmospheric chemist and of sedimentary processes in a way that will pass muster with experts in the science of mud and ooze and yet might engage the attention of a geneticist. I place my hopes for success in this regard less on any ability to make the difficult simple (the difficulties probably being related to the jargons of the disciplines rather than to intellectual profundities) as on the hope that I have sustained an argument that will keep the reader reading.

James Lovelock (and many other "Gaians") will perhaps disagree with much of that argument, but I hope my essay will be read as fundamentally friendly. However, I recognize that domesticating a wild idea is problematic, and my attempt to suggest that Gaia may yet be incorporated into mainstream biology might be seen as a fatal embrace. I can only protest that my intent is rather to assist the idea to thrive by giving it a molecular biological context that will prove as illuminating for Gaia as it has been for more conventionally recognized organisms. For Gaians and non-Gaians and for the uncommitted I hope there is sufficient here to provide an introduction to planetary ecology and to bring together some scattered issues of global habitability and molecular adaptation.

A Word About Units

This book is about the relationships between processes on a wide range of scales in time and space. Thus, the dimensions used must range from those appropriate to describe the history of the planet to those appropriate in the discussion of events at the molecular level. And, therefore, we may need the full range of prefixes and their symbols to characterize the units used, as follows:

Prefix	Meaning
P (peta-)	10^{15}
T (tera-)	10^{12}
G (giga-)	10^{9}
M (mega-)	10^{6}
K (kilo-)	10^{3}
m (milli-)	10^{-3}
µ (micro-)	10^{-6}

n (nano-) 10^{-9}
p (pico-) 10^{-12}
f (femto-) 10^{-15}

Amounts of substances are given in atomic or molar units. This is still not common in environmental texts and it does lead to some oddities. Figure 6.4 may be one of the first times that the application of fertilizer to a corn field has been calculated in kilomoles per hectare! But it is essential to my attempt to relate events at different scales and to understand the interaction of the cycles of different elements. The conversions are easy. For example, industrial activity is adding 6 Gt (6 Pg or 6×10^{15} g) of carbon to the atmosphere annually. Simply divide by 12, the atomic weight of carbon, to give 0.5 Pmoles C per year.

1

The Question of Habitability

That we, author and reader, live on a habitable planet is a truism—evidenced in a most immediate sense by the act of my writing these words and by the presumption of a readership. But, obviously, the planet provides habitat for a multitude of species besides *Homo sapiens*. Less well known is that Earth is looking more and more habitable all the time. That is, we keep discovering that parts of Earth's surface and crust thought to be inhospitable to life are in fact inhabited.

Only recently did we learn, for example, that life has found its way into exposed rock in the frozen deserts of Antarctica. Life finds a home in acidic hot springs and along deep-sea volcanic vents. Even crude oil deposits far below the surface are home to heat-adapted, petroleum-consuming bacteria. Perhaps we should not be surprised by the geographical ubiquity of life forms. The ability of living organisms to inhabit such a wide range of environments has a corollary in the ability of life forms to persist throughout long periods of Earth's history. We shall consider the evidence for the longevity of life on this planet and the implications of that persistence in chapter 3. For now, it suffices to note that geochemical evidence suggests that there have been living systems on Earth for about 3.8 thousand million years (Gy). This means that Earth has been a life-sustaining planet for more than 80% of its existence. Fossil evidence of life appears in the rock record remarkably soon after the end of the last heavy bombardment of meteors. Before the end of this onslaught from space, Earth's surface would have been repeatedly sterilized.

By contrast, all available evidence suggests that Venus and Mars are lifeless now and may always have been so. Yet within the vast range of possible cosmic objects, Venus and Earth and Mars all appear remarkably similar. This is not the place to discuss the much debated question of the prevalence or absence of life elsewhere in the universe. The apparent absence of life forms on Venus and Mars does however suggest that only a narrow range of planetary conditions will permit the genesis and persistence of life (at least as we recognize it). If this is the case, then two questions ensue: What is it about Earth that makes this planet hospitable to life?—or, more specifically, to life based on the chemistry of carbon? Second, will Earth continue to be habitable?—or, perhaps we should ask, how long will it be before Earth becomes uninhabitable for carbon-based life forms?

The second of these questions may be rephrased as a question about the stability of the global environment. I shall not at this point try for a formal and complete definition of "stability" as it applies to Earth's history. Much of chapter 3 can be read as an attempt at such a definition. For our immediate purposes the commonsense usage—a general absence of change or a tendency to resist change—will suffice. Stability is not a sufficient condition for habitability. It is not difficult to imagine planetary environments which, though stable, are inimical to life. The thermal and chemical conditions of the lifeless planets, Venus and Mars, may be stable. If stability is thus not a sufficient condition, it is surely a necessary condition for the persistence of life. There must be limits to the variation of the character of the global environment over time. The long persistence of life over so much of Earth's history implies that whatever may be the environmental limits of life, these have not been breached for at least 3.8 Gy. Proponents of the Gaia hypothesis would point to this persistence as evidence for environmental homeostasis.

Much of this book is concerned with the search for possible mechanisms that preserve and maintain habitability. We will explore how it comes about that the character of Earth's environment has been kept within the bounds of habitability. First, however, we must seek out what is known about the stability or instability of the global environment. The perturbation of the physico-chemical conditions of that environment by human activity, especially over the past hundred years, adds a new dimension to that question. What will be the consequences of continuing such perturbations? How is the global environment responding to the ever increasing pressure of human activity? There is likely no more important question at the intersection of scientific knowledge, public policy, and human concern.

In the next chapter we shall examine, briefly, some of these questions pertaining to human-caused (anthropogenic) impacts. But in this first chapter we shall start on a vastly broader timescale. Notably, what does the geological record tell us about the stability of the global environment within which life forms have evolved over the past 3.8 Gy? Three lines of evidence suggest some considerable measure of physico-chemical instability. First is the evidence of global change implied by mass extinctions. Second is the evidence in the rock record of major shifts in the isotopic composition of inorganic reservoirs participating in the biogeochemical cycles. Third is the evidence of changes in thermal regimes, leading to the "icehouse" climates that gave rise to glaciations. These three lines of evidence that the global environment has been subjected to major perturbing forces—and has been significantly modified by these stresses—will be reviewed in turn.

From the middle of the nineteenth century, geological thought was dominated by the doctrine of *uniformitarianism*, which urged that processes occurring during Earth's history should be examined strictly in the light of the observation of processes occurring on today's Earth. The doctrine of uniformitarianism thus rejected hypotheses requiring the intervention in geological history of catastrophic events with no modern analog. This notion that "the present is the key to the past" (Hutton 1788) has, of course, served geology well by tying the discipline to physics and chemistry. But "the study of Nature such as she now is" (Lyell 1830) left open some of the most obvious of geological questions. One of the principal categories of markers in the geological record over the 580 My (million years) of the Phanerozoic era is provided by relatively abrupt terminations of fossil lineages. These discontinuities are used to denote major stratigraphic boundaries. The classification of geological time thus depends upon *lack* of uniformity.

Extinctions

It is important to distinguish among three ways in which species may disappear from the geological record (Raup 1984). One pattern of extinction might be expected in a relatively unchanging or only slowly or regionally changing physical environment. A species may go extinct in such circumstances of physical stability simply because its biological environment is changing—owing to biological arms races in which nat-

4   The Question of Habitability

ural selection yields more efficient competitors, more effective predators, fewer accessible prey. This pattern of biologically stimulated extinction is happening today—but at a very rapid pace. Because technologically equipped humans are more efficient competitors and more effective predators, species are disappearing at an alarming rate (Wilson 1992). A second way in which species may disappear is by what has been termed *pseudo-extinction*. Here a species becomes so transformed by stepwise evolution that at some point the paleontologist recognizes a new species, although the lineage is unbroken.

Neither of these first two modes of extinction provides much information about the stability or instability of the environment. There is, however, widespread agreement that against this usual background of evolutionary turnover it is possible to distinguish specific periods of "mass extinction." Historical episodes of mass extinction in which a large proportion of species or even higher taxa vanish at or near the same geological time horizon form a third pattern of extinction.

If one plots the number of families of marine vertebrates and invertebrates against time, the progressive increase from the beginning of the Phanerozoic at about 580 Mya (million years ago) to the 750 or so families extant today is interrupted by sharp declines at 440 Mya (–12%), at 370 Mya (–14%), at 245 Mya (–52%), at 210 Mya (–12%), and at 65 Mya (–11%) (Raup and Sepkoski 1982; for a more detailed analysis of these changes, see Benton 1995). These five sharp declines in family-level diversity are widely regarded as mass extinctions. Although paleobiologists agree about the existence of at least five mass extinctions that stand out from the background, there is less agreement about how these anomalies should be interpreted. Do such mass extinctions reflect the cumulative effects of relatively minor quantitative changes in the rate of ongoing processes of speciation and extinction, or are they the consequences of major episodic events?

The answer to that question is important for the relevance of mass extinctions to global environmental stability. On the view that mass extinction events result from a minor increase in the extinction rate combined with a minor decrease in the rate of origination of species (or higher taxa, such as genera or families) (Benton 1985), then there is no qualitative difference between "mass extinctions" and "background extinction," and the former conveys no more information about environmental change than does the latter. If, on the other hand, mass extinctions can be correlated to changes in geochemistry, then such extinctions provide strong clues to the way in which the habitability of

Earth's surface has fluctuated during geological history. Three sorts of nonbiological change may have had profound consequences for the global biota: (1) changes in sea level, perhaps accompanied by changes in salinity; (2) plate tectonic effects, including volcanism; and (3) large body (bolide) impacts. The first two of these possibilities are terrestrially generated, while the third has an extraterrestrial origin.

Although it seems certain that no major changes in ocean chemistry have occurred during the last 600 My (Holland 1984), simple mass balance calculations based on the size of evaporite deposits suggest that from time to time the salinity of the world's oceans could have fallen by as much as 10% (Stevens 1977). Such a change could be sufficiently large to have a severe impact on those species with strict salinity requirements. In the dramatic decrease of marine families at 240 to 250 Mya, species most tolerant of lowered salinity account for a disproportionate share of the survivors. This Permian-Triassic extinction is associated with a major regression of the sea (Holser et al. 1988). There is however no obvious correlation between the pattern of marine evaporite deposition and the pattern of extinction, nor between sea-level patterns and such events (Berger 1984). This does not mean that a factor such as sea level is insignificant. If the causation of extinctions is multifactorial (Erwin 1994), sea level may be a predisposing factor in some circumstances—although, in the absence of other perturbations, it may be without effect on biotic evolution.

It is the third possibility (bolide impact) that has attracted much attention in recent years among scientists and has captured the imagination of the general public. The most striking correlation of a major extinction with a geochemical marker is the finding of anomalous enrichment of heavy elements (notably, iridium) in the very sharply defined sedimentary deposits at or near the paleontologically dated boundary between the Cretaceous and Tertiary (65 Mya). This boundary corresponds to a mass extinction that included, most famously, all the nonavian dinosaurs. A major controversy followed the finding of this geochemical anomaly, particularly the 100–1000 times enrichment in boundary clays. The weight of opinion is now strongly in favor of the original view (Alvarez et al. 1980) that these heavy elements represent the fallout of an impact by an extraterrestrial body. The dating of a 100-kilometer impact crater along the Yucatán Peninsula on the eastern coast of Mexico (Swisher et al. 1992) played a decisive role in the debate. There is still, however, a minority view that iridium enrichment could represent a period of intense volcanism (Garwin 1988). The case for volcanism as a cause of mass extinction has been enhanced by recent close dating of enormous basalt flows in Siberia and India to paleontological boundaries that mark the end-Permian and end-Cretaceous extinctions, respectively (see table 8.1 of Erwin 1993).

The scenario for mass mortality is much the same for either impact or volcanism. Two likely causes of mortality are common to both: (1) air and water contamination by nitric oxide and other toxins, and (2) sunlight-blocking particulates blown into the atmosphere. The scenario is particularly striking for the impact hypothesis. High temperatures generated at the surface of a large bolide entering the atmosphere would generate large amounts of nitric oxide, NO, which could cause a major perturbation to the stratospheric ozone budget (as will be discussed in chapter 2). One estimate of the ozone destruction that is thought to have accompanied the creation of the Yucatán impact crater is as high as 90% (Toon 1984). A 45% depletion of ozone would increase the biologically active ultraviolet radiation at ground level threefold. Such a loading of the atmosphere with NO would also be toxic to the biota, either directly or by effects on the pH of surface waters.

The sunlight-blocking effects of a bolide impact or massive volcanism are also well supported. The impact (or the volcanic activity) would introduce large amounts of dust high into the atmosphere. For some months the dust cloud would envelop the planet. Such a dust cloud would not only curtail photosynthesis but would cause global temperatures to fall dramatically. Quantification of the temperature effect is difficult because the magnitude would depend markedly on the assumed size distribution of dust particles, the seasonal timing, and other factors. Wolfe (1991) has obtained paleobotanical evidence for an "impact winter" that struck the Northern hemisphere at the end-Cretaceous during the growing season—a time when plants would have been particularly vulnerable to a brief but drastic drop in temperatures (25–30°C). Some of the elements ejected by a large bolide collision (and which are anomalously enriched in the boundary clay) are toxic, and surface waters would not only be acidic but polluted (Hsü 1984). Finally, thanks in large part to the scientific bonanza brought by the 1994 comet crash into Jupiter, it is now possible to envision an additional killing scenario for bolide impact: sudden death by heat. The huge volume of material bigger than dust ejected at the impact site would rain back upon the earth in a fire caused by friction against the atmosphere, setting off forest fires perhaps worldwide (Steel 1995).

The biological consequences of all these physico-chemical changes at the Cretaceous/Tertiary boundary are thought to have been extreme. Photosynthetic production in the ocean may have ceased for some period of time (Hsü and McKenzie 1985; Zachos et al. 1989). No land vertebrates heavier than about 25 kg survived into the Tertiary (Hsü 1984), and many smaller animal taxa also vanished. Plants, at least in some geographic settings, were markedly affected (Johnson et al. 1989), spermatophytes giving way to ferns and mosses (Tschudy et al. 1984; Wolfe and Upchurch 1986).

The phenomenon of mass extinctions, whatever their cause or causes, suggests that the habitability of the global biosphere (at least for higher plants and animals) has been threatened more than once in Earth's history. It is far from clear how close to collapse the system has been on such occasions. Nor, until recently, has there been any sense of the nature and magnitude of the disturbing forces. One of the significant intellectual advances that has accrued from the current study of spasmodic events in earth history is not so much the overthrow of uniformitarianism as the prospect of measuring the strength of the forces necessary to disrupt the temporal continuity that was assumed in that doctrine.

Geochemical Shifts

In addition to the fossil record of mass extinctions, a second line of evidence suggests major perturbations of the global environment. This line of evidence pertains to temporal shifts in the isotopic composition of inorganic reservoirs of important elements involved in the biogeochemical cycles. We have already noted that there is a large body of evidence suggesting no major fluctuations in the composition of seawater throughout the Phanerozoic. An important exception seems to be the isotopic composition of sulfates and carbonates deposited in ocean sediments (Holland 1984).

These variations are shown in figure 1.1. A major feature evident in this figure is the relative rapidity of some of the changes, especially considering the sizes of the inorganic reservoirs of carbon and sulfur. Some changes seem to have taken place in less than 0.5 My.

The process responsible for variation in the isotopic content of evaporite deposits of sulfate (^{34}S v. ^{32}S) is certainly the marked isotopic fractionation that occurs during the bacterial reduction of sulfate (SO_4^{2-}) to sulfide (S^{2-}). Bacteria preferentially use molecules of sulfate that contain the lighter isotope of sulfur (^{32}S) rather than the heavier isotope (^{34}S). Thus, the sulfide produced by the process is depleted in the heavier iso-

8   *The Question of Habitability*

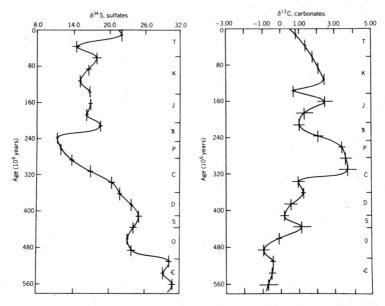

FIGURE 1.1 Instability in Earth's biogeochemical cycles. Note that shifts in the isotopic composition of global carbonates (^{13}C v. ^{12}C) and sulfates (^{34}S v. ^{32}S) are correlated. This correlation may have stabilized concentrations of atmospheric oxygen.
SOURCE: Figures 4.7 and 4.8 of Holser et al. 1988. Copyright © John Wiley & Sons, Inc. 1988. Reprinted by permission of John Wiley & Sons, Inc.

tope. The sulfate that remains will be correspondingly enriched in ^{34}S. An *increase* in the proportion of ^{34}S v. ^{32}S in marine evaporites is therefore evidence of a period of active biological reduction of sulfate.

When the isotopically light sulfide produced in this manner is reoxidized by O_2, the sulfate so formed will be correspondingly depleted in ^{34}S. Thus a *decrease* in the ^{34}S content of evaporites indicates oxidation of sulfide. It might be expected that such major shifts of redox equivalents would be accompanied by large changes in atmospheric O_2. However, figure 1.1 also shows changes in the concentrations of carbon isotopes in carbonate (^{13}C v. ^{12}C) that are negatively correlated with the sulfate ^{34}S changes. These carbon isotope changes are interpreted as changes in photosynthesis, which could counteract the effects of the redox shifts in the sulfur cycle. The increase in the ^{13}C concentration of carbonate is explained as a consequence of increased formation of organic carbon. Photosynthetic organisms preferentially take up carbon dioxide that contains the light isotope of carbon (^{12}C). The carbon dioxide that remains in the atmosphere or ocean and is preserved in carbonate deposits is therefore enriched in the heavy isotope (^{13}C). This interpretation of the carbon isotope record will be discussed in more detail in chapter 3. For now, it is important to know only that the deposition of organic carbon would be accompanied by an increase in atmospheric oxygen. This increase in atmospheric oxygen would thus counteract the consumption of O_2 by the oxidation of sulfides (evidenced by the decrease in sulfate ^{34}S).

Figure 1.1 shows that increases in atmospheric oxygen expected as a consequence of enhanced burial of organic carbon would be offset by decreases in atmospheric oxygen attributable to sulfide oxidation occurring at the same time. This scenario has been quantitatively modeled (Garrels and Lerman 1981). The agreement between the model and the geochemical observations is good. More sophisticated modeling suggests that the offset in the two processes may not have been exact; changes in atmospheric O_2 in the range of 15% to 35% may have occurred (Berner and Canfield 1989). Graham et al. (1995) have emphasized that variation in O_2 of this magnitude could have had a major influence on species diversification in the period between 400 Mya and 200 Mya. Despite such variation, as shall be emphasized in chapter 3, the atmospheric changes must have been confined within limits that permitted the survival of complex multicellular life. The isotope fluctuations do suggest a rough balance between the biogeochemical cycles of sulfur and oxygen that would act to minimize excursions in the oxygen content of the atmosphere. Although a balance of this type may have existed and may have been a major factor in stabilizing the global environmental redox state, it is less clear how the balancing occurs. Notably, what role did the global biota play in bringing about this balance? Nor is it clear what the driving forces are that underlie the isotopic shifts. It has been suggested that the shifts may be correlated with sea-level changes linked indirectly to tectonic events (Holser et al. 1988).

Ice Ages

The third line of evidence suggesting some measure of large-scale instability of the global environment comes from what we know about episodes of glaciation. Evidence of ice ages occurs throughout the geological record, even from the pre-Cambrian. But most attention has been paid to the Pleistocene.

In the last twenty to thirty years it has become possible to make firm inferences concerning temperatures during the Pleistocene. Again, the evidence comes from the different behavior of molecular species, depending on isotopic composition. In this case we are dealing with the behaviors of the molecular species $H_2^{16}O$ and $H_2^{18}O$ (and ^2H_2O, D_2O) in evaporation and precipitation. The lighter species, $H_2^{16}O$, has a higher vapor pressure at a given temperature. Thus liquid water loses $H_2^{16}O$ preferentially to the gas phase when it is subject to evaporation, while water vapor in the atmosphere loses $H_2^{18}O$ preferentially when this gaseous phase is depleted by condensation. These processes, operating on the cycle of evaporation near the equator and condensation in poleward regions, result in polar ice being depleted in ^{18}O and the equatorial oceans being enriched in the heavier isotope. During periods of lower atmospheric temperature and more extended ice sheets, the depletion of ^{18}O in polar ice will be even greater. As ice sheets expand, the $H_2^{18}O$ content of the oceans will increase; as ice sheets contract, $H_2^{16}O$ will be returned to the oceans and the content of the heavier isotope will be correspondingly diluted.

Two observable consequences follow. First, ice cores from Greenland and Antarctica show fluctuations in ^{18}O (or deuterium) content, and these can be used to estimate differences in mean global air surface temperatures. A decrease in ^{18}O content in the ice corresponds to lower temperatures. Second, benthic (seafloor) organisms deposited in the sedimentary column show fluctuations in the ^{18}O content of the calcite (calcium carbonate) in their shells. These fluctuations reflect the ^{18}O content of the water in which these organisms lived. A decrease in ^{18}O content corresponds to the addition of isotopically light water from receding ice sheets (higher temperatures). Higher ambient temperatures also affect the incorporation of ^{18}O into $CaCO_3$, and so for quantitative modeling it may be important to distinguish between the direct effects of temperature and those mediated by the melting of ice sheets (Chappell and Shackleton 1986; Mix and Pisias 1988). Examples of the results obtained from ice cores (in this case, based on deuterium rather than ^{18}O) and from sedimentary calcite are given in figure 1.2.

There is currently widespread agreement that the driving force for periodic glaciation is generated by periodicities in Earth's orbit. The planetary mechanics of diurnal rotation and the annual orbital cycle are responsible for the major changes in insolation received by the Earth, but much slower cycles of the precession of the planetary axis (21 Ky and 23 Ky), the tilt angle (obliquity) of the rotation axis (41 Ky), and the eccentricity of the orbit (100 Ky and 413 Ky) cause fluctuations in the

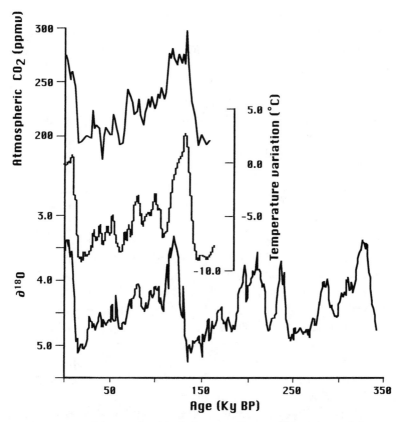

FIGURE 1.2 Instability in Earth's thermostat. *Bottom:* ^{18}O content of the skeletons of *Uvigerina senticosa*, a benthic foram, from an equatorial Pacific deep-sea sediment core—a proxy for global ice volume and temperature (data from Shackleton and Pisias 1985). *Middle:* Variations from the present mean annual temperature at Vostok, Antarctica, deduced from the deuterium profile of the Vostok ice core (data from Jouzel et al. 1994). *Top:* CO_2 content of air entrapped in the Vostok ice core (data from Barnola et al. 1987).

mean annual insolation between minima about 4% below and maxima about 6% above today's value (Hays et al. 1976). Correlation in time between these "Milankovich cycles" (named after the Russian scientist who proposed the relevance of these planetary cycles to climatology) and the occurrence of glaciation and deglaciation is observed, but the changes in solar input are not large enough to account for the changes in temperature inferred from the isotopic evidence. Therefore, although the orbital changes may force the climatic cycle, some amplification mechanism is necessary. Some degree of positive feedback surely must occur from the growth and shrinkage of the ice sheets themselves, which affect planetary albedo, or reflectivity. But it now appears that a most important factor is the greenhouse effect, arising from changes in atmospheric CO_2 and, possibly, other gases (Chappelaz et al. 1990).

Figure 1.2 demonstrates that the content of CO_2 in air trapped in polar ice cores (Barnola et al. 1987) fluctuates in close agreement with the isotope changes that are proxies for the mean global temperature. Periods of low temperature coincide with a low content of CO_2. At the termination of the last glacial, CO_2 rises rapidly from a value between 180 and 200 ppmv to values of 269 to 300 ppmv, characteristic of the preindustrial level of atmospheric CO_2. Relatively small temperature changes recorded by ^{18}O changes during glacial and interglacial periods are also faithfully recorded in the CO_2 record. Changes in atmospheric CO_2 can be inferred also from studies in the difference in ^{13}C content between the remains of surface-dwelling organisms and benthic organisms found in deep-sea sediments. This difference reflects the difference in ^{13}C content between surface water and deep water brought about by the photosynthetic phytoplankton. The photosynthetic process discriminates against the heavier isotope, leaving surface water enriched in ^{13}C. The downward fall of detrital organic carbon from surface waters into deep water subsequently regenerates CO_2 depleted in ^{13}C. It is therefore possible to use the difference in ^{13}C content between the remains of organisms that acquired their carbon from surface water and those that built their shells in deep water to reconstruct the biologically determined chemistry of ancient surface ocean water.

Because atmospheric CO_2 is in equilibrium with (and, on the relevant timescale, determined by) the inorganic carbon of surface water, it is possible to infer historical values of atmospheric concentration of carbon dioxide (pCO_2) from these ^{13}C differences. Because such calculations agree with the direct measurements of CO_2 in ice cores, we can confidently use them to calculate atmospheric conditions back to times for which direct ice core data are lacking. Measurements of ^{18}O in the carbonates of the same sedimentary cores permit comparison of CO_2 changes with changes in the size of the polar ice sheets. Statistical examination of the changes led Shackleton and Pisias (1985) to conclude that, "over all orbital forcing periods, changes in atmospheric CO_2 lead changes in ice volume; that is, these changes must be regarded as contributing to the cryosphere component of the climate record, rather than acting as a passive amplifier."

Whatever the cause(s) of the Pleistocene ice ages, the effects on global biota are striking. Most changes were ecological, however, rather than evolutionary. Major vegetation changes occurred, for example, in Europe, where an alternation is observed between vegetation characteristic of steppe and tundra during glacial periods and temperate forests in the interglacials (Kurtén 1972). In South America, during what appears to have been a dry period between 21 Kya and 14 Kya, savanna vegetation characterized areas which, during warm periods, were occupied by tropical vegetation; open herbaceous growth occurred at the expense of stands of forest (Van der Hammen 1988). As the climate fluctuated, animals either migrated or adapted, as there appear to have been few extinctions associated with the onset of cold periods.

The periodic fluctuations of global temperature during the Pleistocene do not appear to have had the major impact on global habitability as had the events, collisional or otherwise, that wrought mass extinctions earlier in the Phanerozoic. Total global biomass may have been markedly reduced during glaciations (we shall return to this point in chapter 3), but recolonization after glacial termination would rapidly restore the area of closed forest. An interesting exception to these generalizations is among the mammalian fauna, particularly those of large body size. It has been estimated that the average half-life of mammalian species in the Tertiary was 2 My, but that this fell to 1.5 My in the early Pleistocene and to 0.5 My in the late Pleistocene (Kurtén 1972). A major extinction of megafauna took place about 11 Kya, at the end of the most recent glacial. Among the victims were species that had either survived previous interglacials or had left successor species. This extinction leads us to the issue of the impact of anthropogenic pressures on global habitability, as it is believed that this particular extinction is related to the activities of the human population, either directly, as predators, or indirectly, by destruction of habitat.

Perhaps the most important contribution of the phenomena of periodic glaciation to the understanding of environmental stability or instability lies not so much in the amplitude of the changes as in their rapidity. Both the ^{18}O signal and the CO_2 signal change at a glacial to interglacial transition in a time span of less than a thousand years (Broecker 1987; Dansgaard et al. 1989). Some temperature changes appear to have occurred in a matter of decades. It is not yet clear what geophysical mechanisms were responsible for these transitions—which may have been global in extent (Denton and Hendy 1994)—and for the related "Heinrich events," which are defined by marine sedimentary evidence for massive discharge of icebergs into the North Atlantic (Heinrich 1988; Bond et al.

1993; Broecker 1994). On a theoretical level, the possibility of such rapid switches between relatively stable states suggests that the question of stability or instability in global biogeochemistry should be replaced or supplemented by the question of multiple points of stability and the cause(s) of transitions between them. At a practical level, these recent findings suggest that the planetary environment may exhibit instability, not only on the timescale of geological history but on the timescale of human affairs. Such instability is of particular concern at a time in human history when our species is placing unprecedented pressure on the global environment.

Biogeochemical Cycles

The next chapter will summarize some of the ways in which the human species is stressing the pathways of the biogeochemical cycles. Before attempting such an estimate of the global impact of human activity, it will be useful to have a scheme that provides a general description of the flows of material through the soils, water, air, and biota of the planet Earth. These flows are cyclic. Elemental constituents of the air and water are incorporated into living organisms. By excretion, respiration, or upon the death of the organisms or their predators, the elements so incorporated are returned to the inorganic reservoirs. These patterns of flow are the biogeochemical cycles. One of the tasks of this book is to inquire about the role of the biota in the organization and integration of these cycles. Do the biota control cycles that are important for the stability and habitability of the global environment? Before such a question can be approached it is necessary to have an overview both of the global biogeochemical cycles and of the nature and extent of human perturbation.

It would be foolhardy to attempt to review the details of the global biogeochemical cycles in a relatively few pages. Such is not my purpose. There are now a number of edited volumes and monographs, such as *Perspectives on Biogeochemistry* (Degens 1989) and *Biogeochemistry* (Schlesinger 1991), plus a targeted journal, *Global Biogeochemical Cycles*. A series of SCOPE volumes published by the Scientific Committee on Problems of the Environment, a committee of the International Council of Scientific Unions, also explores biogeochemical cycles. Frequent reference will be made to these and other publications throughout this text, which could never have been written without this extensive literature to rely on. Such a large body of secondary sources, not to mention the myriad primary sources on which they are based, presents

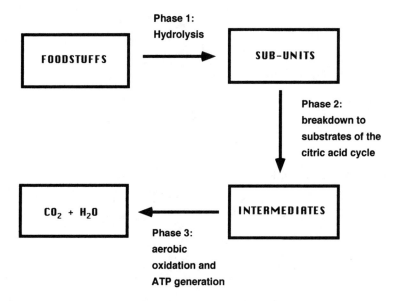

FIGURE 1.3 Cellular pathways and cycles as a model for global biogeochemistry. The complex processes of the catabolism of foodstuffs can be aggregated to a simple four-box scheme. (The terminology used is that of Krebs and Kornberg 1957.)

a difficult challenge for any attempt to reduce current knowledge of the biogeochemical cycles to a single chapter.

A similar difficulty would seem to arise in attempting to understand the breadth and depth of our knowledge of the chemical reactions and translocations that go on in cells and tissues. Biochemists and physiologists view the cycles and pathways of cellular metabolism as orderly in two senses. The intracellular cycles and pathways are made orderly in the sense of being regulated by genetically determined feedback loops, a theme that will become dominant later in this book. They are also seen as orderly in the sense that the multifarious patterns of transformations and transport processes in a wide variety of species may be viewed as variants of a simple ground plan. Any textbook of biochemistry will illustrate this point. Figure 1.3, for example, is typical of such representations of cellular energy metabolism. The orderliness of cellular metabolism is thus discerned by an appropriate conceptual analysis of a field that would otherwise appear as a collection of biological idiosyncrasies. Is it possible to reduce the complexities of the biogeochemical cycles to a simple pattern?

16 *The Question of Habitability*

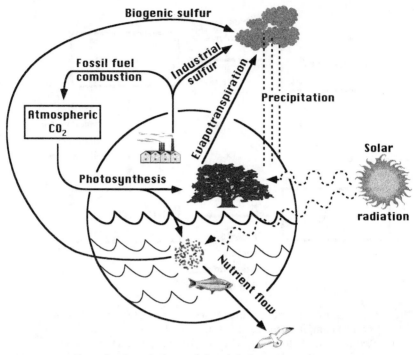

FIGURE 1.4 Complex interactions of the global ecosystem.

The biogeochemical cycles of any ecosystem—local, regional, or global—are commonly represented as systems of interconnected compartments. The number of compartments and the pathways between them will vary for the different elements and for different ecosystems. Any ecosystem within which biogeochemical cycling takes place is almost always complex, made up of interacting populations of different biological species and a diverse physico-chemical environment (figure 1.4). The boundaries of an ecosystem are arbitrary, subject only to the constraint, for biogeochemical studies, that flow rates of the element under consideration must be small across the boundaries compared to those within the system. Such demarcation permits nested components (e.g., a log in a lake in a valley), each one of which may itself for other purposes be treated as a relatively isolated ecosystem with its own characteristic pattern of biogeochemical cycling. Can these manifold characteristics be meaningfully reduced to a simple scheme?

Any attempt to make such a simplification runs counter to current trends in studies of biogeochemical cycles. Interest in the ecosystem as a subject of inquiry arose contemporaneously with the development of

The Question of Habitability 17

powerful methods of large-scale computation. From the formative years of ecosystem theory, researchers were able, with the aid of computers, to attempt the simulation of multiply interconnected, elaborate assemblages of "compartments." The models on which such simulations were based came to demonstrate much of the complexity of the real world. The greater the complexity of the model, the more highly it was regarded. For example, the editor of a collection of classic papers on biogeochemical cycles of the essential elements makes the point that what is needed is, "a clear and correct compartmental model of the system under consideration. If it is a food web, then all significant populations and pathways must be known if the model is to have utility for understanding the real world" (Pomeroy 1974).

As computers have become more powerful, so the models of ecosystem cycles have become more complex. One can now take account of more populations and include increasing numbers of pathways. A fairly typical model of mid-1970s vintage might include ten to twenty compartments (figure 1.5). But by the mid 1980s, models were appearing with thousands of interconnected cells, as in general circulation models of the atmosphere (Heimann et al. 1986). The usefulness of such complex models in the simulation of real world ecosystems has its limitations, however. In the limit they may come to exemplify the paradox of a 1:1 scale map that replicates the complexity of the real world rather than serving as a guide to it.

For the purposes here of achieving an overview of global biogeochemical cycles, the natural complexity will be reduced to a simple four-box scheme, presented in figure 1.6. Formally similar schemes are not uncommon in the literature of ecology (Bormann and Likens 1979), geochemistry (Lasaga 1980), and oceanography (Garrels and Perry 1974). In figure 1.6, however, the four boxes are not defined as chemical or biological species but rather by their functional relationship within the pattern. Similarly, the processes by which the components interact are defined not by their biochemistry, their geochemistry, or their physics but by their role in the ecosystem.

A scheme such as that of figure 1.6 is not intended to simulate any particular cycle, but the aggregation of pathways and compartments serves to identify the common features of all cycles. The scheme is closed, which is appropriate for a global cycle that is self-contained with respect to material, the only input being solar radiation, ultimately reappearing as heat.

Let us walk through the scheme of figure 1.6 to explore how one

18   The Question of Habitability

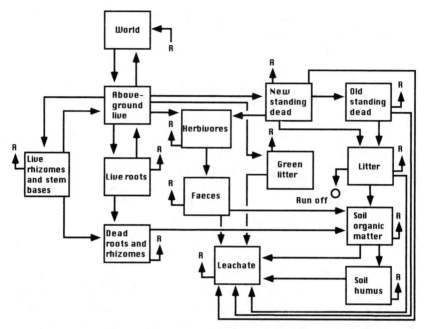

FIGURE 1.5 An early model of ecosystem pathways, with limited complexity. This ABISKO II model is a multiply-interconnected, fourteen-compartment model used to simulate tundra ecosystems. (R represents respiration.)
SOURCE: After figure 1 of Bunnell and Scoular 1975. Copyright © Swedish Natural Research Council 1975. Reprinted by permission of the Swedish Natural Research Council.

might think of the various compartments and the interconnecting processes. Perhaps the best place to start is with the process of *assimilation*. If the element under consideration is carbon, then the global *nutrient* pool would be an aggregate of atmospheric carbon dioxide and the dissolved inorganic carbon of the surface ocean (carbon dioxide plus the bicarbonate and carbonate with which it is in equilibrium). It is upon this nutrient pool of assimilable forms of inorganic carbon that the global autotrophic biota draw, using energy (in most cases, that of sunlight) to reduce the carbon dioxide to the organic carbon constituents of global *biomass*. The much larger pool of inorganic carbon in the deep ocean is made up of the same chemical species as that of surface waters, but it should be put into the *inorganic* box because it is not accessible to life. In the ocean depths, molecules of carbon dioxide (and its hydrated forms) are simply unavailable to the global biota; they are only made available by the upwelling of deep water, identified in the scheme by the arrow labeled *mobilization*.

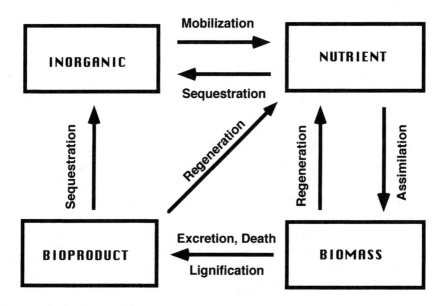

FIGURE 1.6 A simple four-box scheme that provides a framework within which many questions about ecosystems can be asked.
SOURCE: After Williams 1987.

But if one were interested in subsets of the global biota, biomass could be identified with the terrestrial biota and mobilization with the movement of carbon dioxide from the surface ocean to the atmosphere. Conversely, the process of dissolution of carbon dioxide in the surface ocean would be one of *sequestration*, making the carbon unavailable as nutrient. Carbon dioxide is returned to the nutrient pool by the respiration of the primary producers and by the respiration of heterotrophic consumers. In the four-box scheme, both these processes are labeled as *regeneration*, the vertical arrow leading from biomass to nutrient and the diagonal arrow leading from *bioproduct* to nutrient.

For some considerations of the carbon cycle, the nutrient box and inorganic box are chemically identical. But that need not be the case. Consider the case of fossil forms of carbon, much discussed nowadays in connection with the extensive human use of carbon-based fuels. Fossil carbon might be thought of as part of the inorganic pool, or one might put it into the box labeled bioproduct. In the former case, the burning of petroleum hydrocarbons and coal would augment mobilization; in the latter, the human impact would be upon regeneration. Note that the total input into the nutrient pool is always the sum of regeneration (directly from biomass or indirectly through bioproduct) plus mobilization, so the dynamics of the nutrient pool are unaffected by the partition of these processes. The purposes of the modeler will dictate the strategy. In the next chapter I want to group the extraction and burning of fossil fuels with the mining of phosphate and sulfate minerals, so it seems easier to subsume all three of these industrial processes under the heading of mobilization. A primary reason for forcing diverse systems into isomorphous models is that it makes such comparisons possible.

On the other hand, the burning of wood brings about regeneration of nutrient carbon dioxide. But from which pool? The wood of standing forests would usually be thought of as part of the biomass pool, harvested wood being considered as bioproduct. The two pools will have differing kinetic properties, so both would have to be involved. Nevertheless, I shall argue later on that, although it departs from conventional usage, there are occasions when it is better to put both the standing wood and harvested wood into the bioproduct box, reserving the biomass compartment for the actively metabolizing cellular part of the global biota.

The lack of chemical constraints on what is put into a given box, and the corresponding lack of chemical definition of the processes indicated by the connecting arrows, are what gives the scheme generality and, thus, utility. The versatility of the scheme and its usefulness as a flexible framework for thinking about biogeochemical cycling will become apparent as instances of its use are encountered throughout the rest of this book.

Assimilation in the Carbon Cycle

It is the input of solar energy that drives the process of assimilation in figure 1.6. Assimilation is the flow from the nutrient compartment to the biomass compartment of the box model. The primary biochemical process in assimilation is the reduction of carbon dioxide to organic forms of carbon. The first step in this process is the incorporation of CO_2 into 3-phosphoglycerate by reaction with the five-carbon sugar, ribulose-1,5-bisphosphate (RuBP in figure 1.7). This first step is catalyzed by an enzyme which is probably the single most abundant protein in the biosphere, ribulose bisphosphate carboxylase. The reductive photosyn-

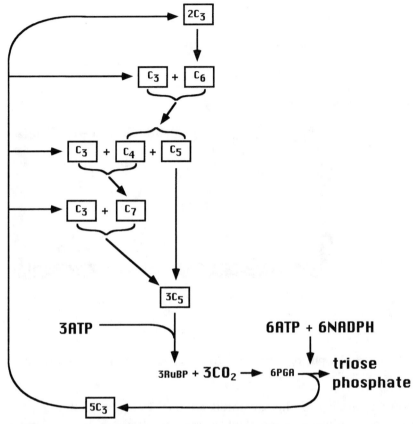

FIGURE 1.7 The Calvin cycle for the net incorporation of CO_2 into carbohydrate. Besides CO_2, the other inputs are ATP and the reductant, NADPH—these are provided by the processes shown in figure 1.8.

thetic cycle that results in the production of one molecule of triose phosphate (PGA in figure 1.7) from three molecules of CO_2 requires the free energy of hydrolysis of the phosphate anhydride bonds of nine molecules of adenosine triphosphate, ATP; six molecules of reduced nicotinamide adenine dinucleotide phosphate (NADPH) serve as reductant.

The ATP and NADPH must be regenerated. This is made possible by the reducing power that arises from the photolysis of water. ATP is resynthesized from ADP and inorganic phosphate at the expense of a chemiosmotic gradient. NADP is reduced to NADPH by ferredoxin, the first stable reduction product of photosystem I (figure 1.8). Importantly for the global biogeochemical cycles, the other product of the photolytic breakdown of water is oxygen. The triose sugar molecules that are the

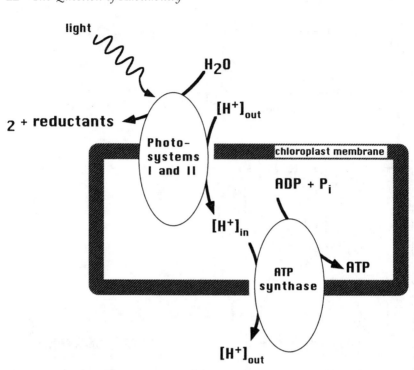

FIGURE 1.8 The photogeneration of ATP and reductants in chloroplasts; oxygen is a by-product.

primary product of photosynthesis are then transformed through the metabolic pathways of the plant or other autotroph into sugars, polysaccharides, lipids, proteins, nucleic acids, and the many other carbon-containing compounds that the biota are capable of synthesizing. All this material is included in the biomass box depicted in figure 1.6. The accumulation of biomass in an ecosystem is called primary production.

It is usual for ecologists to distinguish between net primary production (NPP) and gross primary production (GPP). The difference between NPP and GPP is attributable to the respiration of the photosynthesizing plants (Whittaker et al. 1975). In the four-box scheme these respiratory processes are represented by the arrow from biomass to nutrient. The rate of assimilation is therefore equal to GPP. The biochemical processes just described are those of gross primary production. Gross primary production is not however the same as gross photosynthetic CO_2 fixation. The protein that catalyzes the carboxylation of ribulose-1,5-bisphosphate also catalyzes a reaction with O_2 to give phosphoglycolate. This process requires the light-driven production of the pentose substrate and thus constitutes the first step in photorespiration. The competition between CO_2 and O_2 at this step is of considerable importance with respect to the part that photosynthesis might play in the disposition of anthropogenically produced CO_2. (This matter will be revisited in chapter 3.) For the purposes of the four-box scheme, assimilation (GPP) is equal to gross photosynthetic CO_2 fixation minus photorespiration.

A number of possible factors may limit the rate of assimilation,—that is, the flow of material from nutrient to biomass. In the context of our discussion of biogeochemical cycling, attention will be drawn chiefly to the effects of nutrient limitation, but it is important to remember that terrestrial productivity (production per area or production per biomass) is often limited more by temperature, water, or sunlight. It is thus possible to construct maps of terrestrial productivity on the basis of climatic indices of temperature and moisture (Walter 1973). The construction of such maps is an essential step in attempting to estimate the rate of assimilation in a global scheme.

Estimates of the rate of assimilation exhibit considerable uncertainty. The problem is a difficult one. The estimate is made by summing the production of the various biomes of Earth's surface (Atjay et al. 1979). Three problems are inherent in this approach. First, the assignment of any given region of Earth's surface to a particular ecosystem type can be somewhat arbitrary. Second, estimates of the productivity of each ecosystem type are themselves uncertain. Third, and unsurprisingly in view of the first problem, estimates of the areal extent of each ecosystem type lack a consensus. The production of a given ecosystem type is calculated by multiplying the estimated productivity by the estimated area; global production is arrived at by summing the figures thus obtained. Production is estimated from measurements of the growth of plants in research plots, and it therefore represents NPP. GPP is estimated to be about twice NPP (Bramryd 1979), but there is considerable uncertainty about this ratio. (For an idea of the range of estimates see table 9.3 in Zelitch 1971.) Charles-Edwards (1981) gives a value of 0.4 for the ratio of the daily loss of carbon by respiration to the "photosynthetic integral." From this value we calculate GPP/NPP=1.66. Note, however, that the ratio was determined for crop plants in which the ratio of photosynthetic tissue to nonphotosynthetic tissue would be relatively high.

In summary, all three uncertainties (assignment of ecosystem type, measurement of productivity for that type, and global extent of that biome) affect any estimate of the terrestrial rate of carbon assimilation. Remote sensing by satellite may, in the future, provide new ways of arriving at a better estimate of this rate (Warrick et al. 1986). For now we must live with the quantitative uncertainties. Because the biogeochemical cycles of all the other elements incorporated into biomass are driven by the photosynthetic incorporation of carbon, the rate of carbon assimilation is fundamental for the quantitative study of the global biogeochemical cycles.

Bolin et al. (1979) estimated terrestrial NPP to be 4.42 Pmoles/y (Peta-moles, or 10^{15} moles, per year). A corresponding estimate of GPP is arrived at by multiplying this figure by 1.75—though, as just discussed, the adoption of any such conversion factor for transforming NPP into GPP is fraught with uncertainty. To arrive at an estimate of global assimilation, we now must add in marine production. Even greater uncertainty attaches to the figures for marine production than for terrestrial production. Problems in the methodology for measuring production and the problems of defining regions of productivity and of assessing the areal extent of those regions produce a considerable range of estimates. As in the case of terrestrial production, it is likely that the near future will bring rapid advances in the use of remote sensing to provide better estimates of marine assimilation of carbon (Sathyendranath et al. 1991). A marine GPP estimate of 3.63 Pmoles/y was made by De Vooys (1979). This figure can be added to the estimate for terrestrial GPP to give a global rate of carbon assimilation of 11.36 Pmoles/y. Given all the combined uncertainties, however, there is no justification for reporting this value to four significant figures.

This global assimilation of carbon from CO_2 into carbohydrate will be accompanied by the liberation of a stoichiometrically equivalent amount of O_2. The reaction

$$6CO_2 + 6H_2O \rightarrow C_6H_{12}O_6 + 6O_2$$

is accompanied by a free energy change of +2833 KJ/mole (kilo-joules per mole). The free energy change associated with global photosynthesis is therefore $(2833/6) \times 11.36 \times 10^{15} = +5.4 \times 10^{18}$ KJ/y. For comparison, the solar energy input to the Earth's surface is 2.9×10^{21} KJ/y, of which 1.3×10^{21} KJ/y is in the photosynthetically active wavelength range of 380 nm to 780 nm (Slatyer 1973). Similar values for the efficiency of fixation of solar radiation are calculated by Lieth (1975).

Regenerating the Nutrient Pool

If the global ecosystem, as represented in the four-box scheme of figure 1.6, is in a steady-state, then the rate of depletion of the nutrient pool by assimilation must be balanced. As the scheme shows, there are three inputs to that pool. One input is the mobilization of biologically inert materials. A second is the regeneration of nutrient from biomass. A third is the regeneration of nutrient from bioproduct. These three inputs differ in the length of the pathways for the return of the products of autotrophic assimilation:

biomass → nutrient

biomass → bioproduct → nutrient

biomass → bioproduct → inorganic → nutrient

In relation to the dynamics of the global cycle, the differences between the three pathways lie in the size and kinetic properties (turnover times, figure 1.9) of the reservoirs intervening between living organisms and the nutrient pool. Because turnover time is calculated as the ratio of reservoir size and flux rate, the generally huge inorganic reservoir will typically have a long turnover time—perhaps on the order of thousands to millions of years. In contrast, the turnover time of the bioproduct pool will be relatively short, measured in tens or hundreds of years. Biomass will have a turnover time of a year or less, with a corresponding half-life ($t_{1/2}$), or response time, of a few months. This short response time for direct regeneration of nutrient from biomass is exemplified by the well known annual upswing of atmospheric CO_2, which is such a marked feature of the Mauna Loa records. The four pools, or compartments, of figure 1.6 are in fact distinguished more by their different turnover times than by their distinctive chemistries. Any attempt at detailed modeling would thus need to represent a series of pools of differing turnover times. In Bolin's 1983 model of the terrestrial carbon cycle, five compartments are distinguished: "short-lived biota" ($t_{1/2} = 0.82$ y), "litter" ($t_{1/2} = 0.76$ y), "long-lived biota" ($t_{1/2} = 32$ y), "peat" ($t_{1/2} = 110$ y), and "soil" ($t_{1/2}$ between 200 y and 500 y).

In the carbon cycle the direct regeneration of CO_2 occurs either by the respiration of primary producers or by the respiration of consumers. Biochemically, these two processes are indistinguishable. Both involve

$$f_{in} \rightarrow \boxed{R} \rightarrow f_{out}$$

$$f_{in} = f_{out} = f$$

turnover time = R/f

half-life = 0.693 R/f

FIGURE 1.9 The relationship between turnover time and half-life in an open steady-state system. Here R signifies the reservoir size and f is the rate of flux. The half-life is the time in which, if the input were cut off ($f_{in} \rightarrow 0$), R would fall to 50% of the steady-state value. It is a mathematical property of the potential decline that half-life = $\ln 2 \cdot R/f$ ($\ln 2 \approx 0.693$).

the production of CO_2 as an end-product of a series of enzymic reactions that abstract hydrogen from reduced organic compounds for oxidation via the mitochondrial electron transport chain; molecular O_2 is the ultimate sink for electrons. Concomitantly, energy is trapped, initially as an electrochemical gradient of protons and, subsequently, as the free energy of hydrolysis of phosphate anhydrides (figure 1.10). Processes of biochemical decomposition similar to those that underlie the direct regeneration of CO_2 via the "dark" respiration of autotrophs and heterotrophic respiration are also responsible for the indirect regeneration of CO_2 from humus and from the long-lasting pool of dead organics within the reservoir denoted in figure 1.6 as "bioproduct." The overall equation for respiration is a reversal of that already given for photosynthesis. During global processes of respiration, the oxygen that was produced by the photosynthetic reactions of global primary production is consumed. Respiration is exergonic, and the energy of solar radiation that was used to drive photosynthesis is made available to respiring autotrophs and consumers.

As we have seen, about half of the carbon that is photosynthetically incorporated into terrestrial biomass (GPP) is returned directly to the atmosphere by the dark respiration of these same photosynthetic

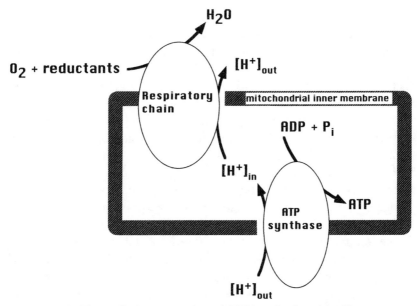

FIGURE 1.10 The oxidative generation of ATP in mitochondria. The oxygen arises from the processes shown in figure 1.8. Note that the direction of the transmembrane (inside ↔ outside) movement of protons (H+) differs between chloroplasts and mitochondria.

organisms. The remaining carbon, which is stored in terrestrial biomass (NPP), is returned to the atmosphere as CO_2 by three principal routes: heterotrophic respiration by macroorganisms, fire, and microbial respiration. None of these three routes can be quantified with accuracy, but some estimates have been made. Of the 4.4 Pmoles of carbon fixed each year by terrestrial photosynthesis, perhaps between 4 and 19% returns to the atmosphere by the respiration of heterotrophs other than microorganisms (Atjay et al. 1979). A significant proportion of annual NPP is returned to the atmosphere by fire, mostly set deliberately for agricultural purposes. Wong (1978) suggested that the atmospheric input of CO_2 from burning wood could amount to as much as 10% of NPP, though Fahnestock (1979) considers this an overestimate by a factor of about 4. The estimate of the flux of carbon by biomass burning by Seiler and Crutzen (1980) is between 0.17 and 0.34 Pmoles/y. Other than heterotrophic respiration and fire, the remainder (about 90% of NPP) is accounted for by microbial decomposition of short-lived bioproduct (litter, about 70%) and longer lasting bioproduct (peat and soil, 5 to 10%) (Bolin 1983).

In the case of marine NPP, more than 95% is returned directly to the surface water by phytoplankton respiration or by the respiration of consumers (Broecker and Peng 1982). The remainder is removed from the photic zone, as detritus falls under the influence of gravity or as net productivity is carried downward by the vertical migration of animals. About a fifth of this sinking carbon is in the form of calcium carbonate, which constitutes the hard parts of some species of phytoplankton and zooplankton (Broecker 1973). The total amount of carbon removed from surface waters is relatively small—at most a few percent of total global GPP—but it is of crucial importance because the undecomposed organic detritus carries not only carbon but other potential nutrients, such as nitrogen and phosphorus. Nitrogen and phosphorus accompany carbon in organic molecules in, roughly, the "Redfield ratios" (to be discussed in chapter 6). In the four-box scheme of figure 1.6, note that the process of sequestration need not necessarily represent a chemical transformation; it may equally well represent a translocation in which the new physical location is unpropitious for assimilation. The loss of nitrogen and phosphorus from the photic zone owing to the fall of detritus is a good example of sequestration by translocation.

Nitrogen and phosphorus are carried into dark, deep waters through the downward drift of a variety of organic products generated in surface waters by photosynthesis. As the organic matter sinks, this "rain of detritus" is subject to bacterial oxidation. The oxidation of these biological residues may be summed up in an equation (Richards 1957; Krumbein and Swart 1983.) This equation is

$$(CH_2O)_{106}(NH_3)_{16}(H_3PO_4) + 138O_2 + 18OH^- \rightarrow 106CO_2 + 16NO_3^- + HPO_4^{2-} + 140H_2O.$$

The observable result of this process is an increase with depth of the concentrations of inorganic carbon, nitrate, and phosphate and a decrease in the concentration of dissolved oxygen. A typical profile of these solutes vs. depth is shown in figure 1.11. The profiles show clearly the photosynthetic production of O_2 at the surface of the ocean and the concomitant depletion of nutrients, and they show the complementarity of the nutrient/depth profile with the oxygen/depth profile that would be anticipated from the equation just presented.

Two important features of ocean biogeochemistry must be noted. First, there are areas of the ocean where oxygen depletion at depth is very marked—indeed there are a number of special areas where anaerobic

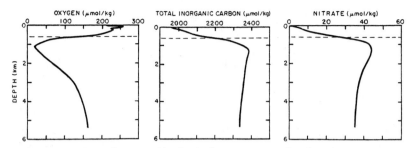

FIGURE 1.11 The influence of biological processes on the chemistry of seawater.
SOURCE: Figures 1.1 and 1.2 of Broecker and Peng 1982. Copyright © Wallace S. Broecker and Tsung-Hung Peng 1982. Reprinted by permission of the authors.

conditions exist. Second, in surface waters phosphate and nitrate are almost completely removed. Any nutrient regeneration is immediately taken into the standing crop of biomass. In a later chapter I will suggest that there is no need to debate which of the two (phosphate or nitrate depletion) is the limiting factor for growth. Because both are present at suboptimal concentrations, the constraint on phytoplankton growth may indeed be a function of both factors.

The existence of zones of low O_2 tension in the sea increases the possibility that organic matter, falling from the photic zone, may in fact reach the sediments and escape aerobic oxidation. Only a small proportion of the primary production of the oceans likely reaches the sea floor in this manner (<1%). This small leak from the carbon cycle, however, is responsible for the net production of photosynthetic oxygen. If this organic carbon did not escape oxidation, the annual cycles of photosynthesis and respiration would be exactly balanced and no photosynthetic oxygen would remain in the atmosphere. If, on the other hand, some of the reduced carbon products of photosynthesis are buried on the ocean floor, then a stoichiometrically equivalent amount of oxygen must remain in the atmosphere. A final balancing is achieved only when the organic carbon is recycled tectonically and returned to the atmosphere as CO_2. The significance of this point will be discussed in more detail in chapter 3.

The loss of nutrients from the photic zone implies that the rate of marine photosynthesis is limited by the rate at which the surface ocean receives new inputs of these nutrients. In our four-box scheme, the process of sequestration can be either chemical conversion to a biologically inert form or physical segregation, as in the case of the gravita-

TABLE 1.1
Kinetics of Phosphorus Pools

	Pool Size (Tmoles)	Flow Rate (Tmoles/y)	Half-Life (years)
Biota	2.7	260	0.073
Surface water	13	2.9*	3.1
Deep water	2568	2.9*	610
Ocean	2581	0.065	27400
Sediments	2.7×10^7	0.065	2.9×10^8

*rate of circulation by upwelling.

tional removal of nutrients. Similarly, the process of mobilization may work either through chemical transformation of a biologically inert form of the element into one which is available for assimilation or through translocation of the element into a zone where photosynthetic activity is possible. The resupply of phosphate to surface marine waters occurs by purely physical processes, either by the input of phosphate from rivers or by the upwelling of deep seawater that has been enriched by falling organic matter. In coastal waters, the riverine input may be significant; but for the ocean as a whole, the input of phosphorus from rivers is only about 1% of that from deep water (Broecker and Peng 1982). Rather, it is the upwelling of enriched subsurface waters that sets the rate of the nutrient-limited biogeochemical cycle of the ocean.

The link between the physics of oceanic circulation and the biogeochemical cycles is thus of primary importance. Nevertheless, the 1% of the input of oceanic phosphorus that is derived from river water must somehow be balanced by removal from the oceans in order to maintain steady-state conditions. As in the case of carbon, some of the phosphorus used in the biogeochemical cycles of surface waters must eventually reach the sediments of the ocean floor and be buried. This sequestration is maintained until the sedimentary deposits along continental shelves are uplifted and subjected to weathering and until the ocean floor is recycled in subduction zones and eventually returned to the surface. Modeling of the phosphorus cycle thus requires a series of pools of widely differing turnover times. Table 1.1 summarizes the kinetics of the principal planetary pools of phosphorus. The existence of such a large range of values permits the modeler to consider subcycles in relative isolation. For instance, a marine biologist concerned with the dynamics of phosphorus cycling in the biota of surface waters can

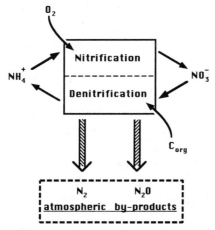

FIGURE 1.12 Nitrification and denitrification. Both processes can generate atmospheric by-products.

legitimately ignore the movement of phosphorus through the sedimentary cycle.

In the nitrogen cycle of the ocean, the process of sequestration can be thought of as the physical removal of material from the site of photosynthetic primary production. Nitrogen is also sequestered from the active biological cycle by a chemical transformation that makes it unavailable to the biota. Nitrogen can occur in valence states from -3 to $+5$. In its most reduced form (NH_4^+, ammonium) and in its most oxidized form (NO_3^-, nitrate) nitrogen is directly available to plants, although NO_3^- must be reduced to NH_3 before it can be incorporated into the nitrogenous compounds of cells and tissues. Nitrate can also be used by a wide variety of microorganisms as an oxidant in anaerobic respiration, and a number of chemautotrophs can use NH_3 as a source of reducing power. These two processes of nitrification and denitrification are interlinked (figure 1.12). Both processes can give rise to N_2 and N_2O, which escape to the atmosphere and which cannot be used by plants.

The rate of sequestration of nitrogen by nitrification and denitrification into forms not directly available to plant photosynthesizers is not insignificant compared to the rate of assimilation. The rate of denitrification by terrestrial systems is about 10 Tatoms of nitrogen (in the form of N_2 or N_2O) per year (Delwiche and Likens 1977). The C/N ratio for the terrestrial biota is 80. Therefore, the value of NPP presented earlier (4.42 Pmoles of carbon per year) is equivalent to 55 Tatoms of nitrogen during the same time. The rate of assimilation of nitrogen will be higher than 55 Tatoms N/y for two reasons. First, recall that assimilation in the four-box scheme is equivalent not to net primary production but to gross primary production (GPP). Second, the biochemical processes underlying GPP are those of the photosynthetic tissues, which may be expected to have C/N ratios not dissimilar to those of marine algae, a point that will be further elaborated in chapter 6 when we explore the connection between the cycles of carbon and nitrogen at the molecular level. (For a preview of the molecular processes of the global nitrogen cycle in the four-box form, look ahead to figure 6.3). The terrestrial rate of nitrogen assimilation may therefore be as high as GPP × 16/106 = 1170 Tatoms/y. Even at this higher value for nitrogen assimilation, the "leakage" rate from the terrestrial nitrogen cycle is at least 1% of the flow rate into the terrestrial biota. It will be important for future research to establish the relative roles of nitrification and denitrification in the production of N_2 and N_2O. The uncertainty in the figures given in figure 1 of Söderlund and Svensson (1976) has not been significantly reduced.

To balance circumstances in which sequestration of any element is brought about by translocation rather than chemical transformation, the reverse processes of mobilization must also be operative as a translocation. Only in this way will the four-box cycle come into balance. If, however, the process by which an element is rendered biologically inert is a chemical transformation, then mobilization will consist of the appropriate chemical reversal of that transformation to make the element available as a nutrient once again. The nitrogen cycle provides an obvious example of the route of chemical transformations. The major proportion (>99.9%) of the nitrogen of the atmosphere and dissolved in the oceans is in the form of dinitrogen, N_2. Few living species (indeed, only certain microbial taxa) are able to utilize N_2. Of the 1170 Tatoms of nitrogen incorporated annually in terrestrial GPP (7.73 Pmoles × 16/106), over 99% will be assimilated from NH_4^+ and NO_3^-. The pool of such nutrient nitrogen is however continuously being depleted by the side reactions of the processes of nitrification and, primarily, denitrification (figures 1.12 and 6.3). The size of the terrestrial pool of soil inorganic nitrogen is about 10^4 Tatoms (Rosswall 1983). If the terrestrial denitrification rate is 10 Tatoms/y, then the half-life of the ammonium (NH_4^+) and nitrate (NO_3^-) of the global soils would be as little as 1000 years. Such a rapid depletion must be counterbalanced by some process of nitrogen mobilization.

The Question of Habitability 33

There are three ways in which N_2 can be converted into assimilable forms. One is biological, another is abiological, and the third route depends on human intervention. First, a limited number of procaryotes (bacteria) are able to utilize N_2 by reducing it to NH_3. The reduction is carried out by an enzyme, nitrogenase, the properties of which will be discussed in detail in chapter 6. (There I will show that this enzyme has properties that make it a prime candidate for a role in the control of the global nitrogen cycle.) Determining the global rate of biological fixation of nitrogen is a difficult endeavor. Estimates of the terrestrial fixation rate vary from 3 to 14 Tatoms/y (table 2.4 of Rosswall 1983). This range does bracket independent estimates of the denitrification rate. Estimates of the marine rate of biological nitrogen fixation and of denitrification are even more widely dispersed. The biological process of nitrogen fixation is reductive, whereas the abiotic pathway is oxidative. Nitrogen oxides are produced abiotically in the atmosphere by electrical discharge in thunderstorms. All estimates place the rate of the abiotic path of nitrogen fixation lower than that of the biological fixation of N_2. Many experts place it considerably lower. On the other hand, the rate of fixation of N_2 by human activity, deliberate or inadvertent, is now approaching the sum of the rates of the first two processes. This major perturbation will be assessed in chapter 2.

As with the nitrogen cycle, the sulfur cycle also provides examples of sequestration and mobilization by chemical transformation. Like nitrogen, sulfur can exist in a number of valence states. Again, as in the case of nitrogen, a number of important metabolites of the global sulfur cycle are gases—sulfur dioxide (SO_2), hydrogen sulfide (H_2S), carbon disulfide (CS_2), carbonyl sulfide (COS), and dimethyl sulfide (($CH_3)_2S$). One of these, dimethyl sulfide, is attracting much current interest and will be discussed in chapter 3. Sulfur also provides an important example of sequestration by mineralization. In anaerobic sediments, SO_4^{2-} is transformed by sulfate-reducing bacteria to H_2S. A variety of processes can then lead to the reoxidation of this H_2S upon its diffusion toward the oxygenated surface. Some fraction, x, of this sulfide will react with metal ions, primarily of iron, to form highly insoluble metallic sulfides. Because most anaerobic sediments are marine, this physical reservoir must be the focus of attention with respect to the global sulfur cycle. The deposition of metallic sulfides in the global marine sediments amounts to about 3.6 Tmoles/y (Ivanov 1983). The total amount of H_2S produced by sulfate reduction in marine sediments is therefore about

3.6/x Tmoles/y. The equation for sulfate reduction, given by Krumbein and Swart (1983), is

$$(CH_2O)_{106}(NH_3)_{16}(H_3PO_4) + 53SO_4^{2-} \rightarrow 106CO_2 + 16NH_3 + H_3PO_4 + 53S^{2-} + 106H_2O.$$

The ratio of carbon oxidized to sulfur reduced is 2:1.

Thus, sulfate respiration would be accompanied by a corresponding oxidation of 7.2/x Tmoles of reduced carbon out of a total marine GPP of 3.63 Pmoles of carbon per year. The equation also shows that this reduction of sulfate is not accompanied by liberation of O_2, so the conclusions reached earlier concerning the relationship between the burial of the reduced products of photosynthesis and the net photosynthetic production of a stoichiometrically equivalent amount of oxygen are not affected. The conclusions would need to be modified only in that the buried reducing material would be in the form of sulfides (such as FeS and FeS_2) rather than organic carbon. The burial rate of organic carbon is about 10 Tmoles/y (table 6.7 of Holland 1978). Each mole of sulfide buried carries twice as many reducing equivalents as does a mole of organic carbon, so it appears that reducing equivalents are being buried in the proportion of 10 (as carbon) to 7 (as metallic sulfides). It would be useful to obtain a global average for the value of x, which is the ratio of sulfide precipitated to the total production of H_2S (i.e., the oxidation of organic carbon by SO_4^{2-}). The value of 0.1 found by Jorgensen (1983) would necessitate the oxidation of 7.2/0.1 = 72 Tmoles of carbon per year. This would represent about 0.2% of marine GPP, so the figure is probably too high. However, the situation in coastal sediments to which Jorgensen's results apply will be such that an unusually low fraction of sulfide will be trapped as metallic sulfides (Holser et al. 1988). A value of x approaching 1 would reduce the fraction of marine GPP being used in sulfate reduction to a more likely number.

We here encounter an example that has general significance for the main thrust of this book. Understanding the global sulfur cycle turns out to depend upon an understanding of the small-scale events taking place in just one kind of locale: marine sediments. I pointed out earlier that fast subcycles can be considered in relative isolation from slow cycles. The converse may not apply. Modeling of the slow geochemical macrocycles can, under some circumstances, depend upon details of the relevant fast biogeochemical subcycles. In chapter 6 I shall conclude that it is possible to extend this principle by suggesting the dependence of global cycles upon subcellular, molecular systems.

All living systems on Earth depend on the effective cycling of key elements on a global basis. These bioavailable elements must cycle between four compartments, or reservoirs: biomass, bioproduct, inorganic, and nutrient. The fluxes are processes of sequestration and mobilization, of assimilation and regeneration. In some cycles these processes may involve physical translocation; in others they are chemical transformations.

This brief overview of the global biogeochemical cycles suggests one or two problems on which a metaphor of global metabolism may cast some light, or which may be reformulated within the context of such a metaphor. The question of how the cycles interact is one such problem because, certainly for nutrient elements, the principal site of interaction is in the metabolism of the biota. A second question concerns the coordination of the rates within a given cycle. In all cases, the rates of assimilation are much higher than the rates of leakage from the biological cycle and the corresponding rates of replenishment of the cycle. That is to say, the biological cycles of C, N, S, and P are relatively closed, and the biota are capable of recycling these elements in a very efficient manner. But the cycles are not completely closed; the outflows to the inorganic pool are of sufficient magnitude that, if there were no corresponding mobilization, the biological cycle would disappear quickly into the geological background. This has not happened. That life has persisted for at least 3.8 Gy is an empirical finding, revealed in the fossils of microbial life. For Barlow and Volk (1990) the persistence of life poses a question in system theory: "The puzzle is this: How can an aggregate of open-system life forms evolve and persist for billions of years within a global system that is largely closed to matter influx and outflow?"

The answer to this general question must include either some mechanism to restrain global environmental conditions within a range that will permit life to persist (chapter 3) and/or the ability of biological adaptation to meet the challenges of the noise and drift in the environmental signal (chapter 5). Gaian metaphors (chapter 4) suggest a major role for the biota in maintaining the rough equality between sequestration and mobilization that is essential to global biogeochemical stability and, therefore, a necessary condition for planetary habitability.

2

Dimensions of the Anthropogenic Perturbation

At this point in human history, the question of whether the global environment is or is not stable is of particular interest, in view of the many ways in which the human population is impacting upon the planetary system. A useful general way to think of the effect of human activities is to reformulate questions concerning the global environment in the form of inquiries about the stability of the parameters that characterize that environment. For instance, expressions of concern or alarm as to what is happening or may happen to the mean temperature of Earth's surface, or to the level of ozone in the atmosphere, presume that these parameters are unstable. On the other hand, those who dismiss or discount such expressions of unease are asserting, in effect, that the parameters under discussion are stable, at least to the current level of anthropogenic pressure. We shall turn shortly to discuss the magnitude of that pressure on the global system. Any discussion of stability must always take into account the dimensions of the perturbation.

As the size of the anthropogenic impact becomes clearer, the question of stability must be revisited. For instance, when Barbara Ward and René Dubos were writing the Preparatory Report for the United Nations Conference that was held in Stockholm in 1972, they were impressed by the confusing spectrum of views forwarded to them by the members of a highly distinguished panel of consultants. They made a list of some of these dichotomous opinions (Ward and Dubos 1972), and at the head of the list appears "some are more impressed by the stability and resilience of ecosystems than by their fragility." Twenty years later, in preparation for the analogous conference in Rio de Janeiro, the rate and extent of the changes that had taken place since the Stockholm meeting were summarized in the UNEP report, *The World Environment 1972–1992: Two Decades of Challenge* (Tolba et al. 1992). It would be interesting to re-poll the 1972 consultants twenty-three years later to determine whether more recent estimates of the magnitude of human impacts would cause them to modify their responses. The widespread public concern about the future of the planet means that, among the major scientific questions of the last decade of the twentieth century, none possesses more human interest than the question of the stability of the global environment. The stability of "Nature" has been the object of human concern throughout much of recorded history (Wiman 1990), but the sources both of predictability and of caprice were until now thought to be inherent in the natural world. Only recently has it been suggested that human activity could generate sufficient pressure to inadvertently shift or reset some of the parameters that characterize the human environment.

It can be argued that the state of scientific knowledge is not yet adequate to provide answers to questions of anthropogenic impacts on the stability of global climate and chemistry. There is, after all, considerable uncertainty about the factors that stabilize or destabilize even local ecosystems, such as lakes or watersheds. Gates (1975) has warned against "bypassing the fundamental ecological issues in a hasty attempt to solve quickly the large-scale ecosystem problems." Certainly, this book is written more with the intention of raising questions than providing answers. But there is an important extrascientific reason for raising the global questions. The bias toward the global issues is, in part, an ethical response to the realization, novel to the twentieth century, that human activities may conceivably be placing such a stress on global systems as to jeopardize the welfare of all people everywhere. Paradoxically, it is also an attempt to withdraw from questions of value.

The despoliation of any particular area of Earth's surface as a result of human activity may always be balanced by gains in human welfare produced by the activity of which the despoliation is an unhappy by-product. What measure of disruption of natural systems is justified by how great an extension or improvement of human welfare is a problem on which people of goodwill may differ (Editorial 1990; Weatherley 1990). For instance, although many would deny this thesis, it appears not totally irrational to insist that even the damage to thousands of square kilometers caused by emissions from smelters are worth the benefits accruing to society from the ready availability of nickel and copper. These problems of environmental costs versus societal benefits take on a new dimension when they are raised in the context of the developing countries (Editorial 1993; Parikh and Painuly 1994).

When we move the question of impacts to a global scale, the difficult problems of responsibility and amelioration remain unaddressed, but at least the ethical questions are somewhat less complex. This is because it is human welfare everywhere that is at risk if there is any marked deterioration in the habitability of the planet. To use a clinical analogy, the questions of global stability or instability concern the point at which a local infection becomes systemic, no longer self-contained, and thus a threat to the life of the organism. At a time when the physico-chemical characteristics of global systems may be shifting, the questions of value become greatly simplified and it may thus be possible to approach problems of the stability of global parameters with less distraction from the competition of regional self-interest or differing cultural value systems than is the case when local perturbations are the object of discussion.

An attempt to recast critical environmental questions in terms of stability and habitability may also assist in removing these questions from the contentious arena of ethics and politics because attention is focused on the properties of the unperturbed system rather than on the indictment of individuals or institutions who, consciously or inadvertently, may stress the system. Conversely, arguments concerning value-laden environmental issues might be somewhat clarified if a clear understanding of the stability and habitability of the global ecosystem could be achieved. In other words, a detached study of the stability of global systems may play a part along with passionate advocacy in the attempt to ameliorate the human environment.

In the previous chapter a simple four-box scheme (figure 1.6) was used to sketch some of the principal features of the global biogeochemical cycles. The four boxes, or compartments, are inorganic (or abiotic), nutrient, biomass, and bioproduct. The interconnecting pathways are also given functional appellations: mobilization, sequestration, assimilation, regeneration, excretion, lignification, and death. In this chapter the four-box portrayal of global reservoirs and the associated fluxes will be used to examine the impact of human activities on the global cycles of four major nutrient elements—carbon, nitrogen, phosphorus, and sulfur—which are the processes that supporters of the Gaia hypothesis might refer to as "global metabolism."

Human Impact on the Carbon Cycle: Atmospheric Carbon Dioxide

The global biogeochemical cycle of carbon is being affected by humans in a number of ways. The net effect of changes in what may be considered the global metabolism of this key element is clearly observable in the records of atmospheric CO_2, the longest and best known of which is from Mauna Loa (figure 2.1), though similar trends are observable in records from other geographical locations in both hemispheres. There is no doubt that the annual average level of CO_2 has increased from 316 ppmv in 1960 to 356 ppmv in 1992. If the volume of the atmosphere (at standard temperature and pressure) is 4.0×10^{21} liters (Verniani 1966), then 1 ppmv of CO_2 corresponds to $4.0 \times 10^{15}/22.4$ moles CO_2, or 2.1 Pg (Peta-grams, 10^{15} grams) of carbon. Thus, over the past three decades, the quantity of CO_2 in the atmosphere has been increasing at a rate of about 0.18 Pmoles/y. The current rate of increase is about 30–40% higher; Bolin (1986) uses a value of 0.22 Pmoles/y for the rate of increase in the mid 1980s, and that value will be adopted in some calculations in this chapter.

Considerable effort has been devoted in recent years to obtain a quantitative account of the changes in carbon flux that account for the increase in atmospheric carbon dioxide. It is important to recognize that one of the chief difficulties in providing such an account arises from the fact that this net flux into the atmosphere represents only a small fraction of the steady-state fluxes to and from the atmosphere. The principal exchanges in and out of the atmosphere are with the surface waters of the ocean and with the biota.

The physical exchange of CO_2 between the atmosphere and the oceans is a complex process, varying with the temperature and with the concentration of dissolved inorganic carbon in the surface water. Both of these factors depend upon latitude. For a given concentration of dissolved inorganic carbon ($[CO_3^{2-}] + [HCO_3^-] + [CO_2]$), the partial pressure of CO_2 in equilibrium will decrease with decreasing temperature. Because the processes of atmospheric mixing are relatively fast and maintain an approximately uniform pCO_2 from pole to equator, CO_2 is transferred from the atmosphere to the ocean at high latitudes. Conversely, at low latitudes, there is a net transfer of CO_2 from the ocean to the atmosphere, a process that is accentuated by the upwelling of waters that have been biologically enriched with CO_2, owing to decomposition of organics falling beneath the photic zone. In the Pacific equa-

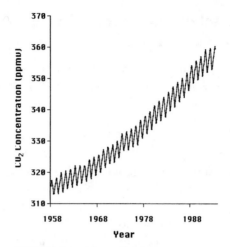

FIGURE 2.1 Atmospheric concentration of CO_2 measured at Mauna Loa (1958–1993).
SOURCE: The data are taken from Keeling and Whorf (1994).

torial waters the disequilibrium between atmosphere and surface ocean can be as high as 80 ppmv (Broecker and Peng 1982:159), which is about 25% of the mean atmospheric pCO_2. The area of upwellings is however restricted, and the net transfer of CO_2 in this way is small compared with the global total of diffusive transfer across the atmosphere-ocean interface.

This transfer can be estimated from measurements of the ^{14}C content of surface water, which, prior to nuclear weapons testing, was about 4% lower than that of atmospheric CO_2 (Broecker and Peng 1982:121). This deficit of ^{14}C, brought about by radioactive decay, must be balanced, in the steady-state, by an uptake of ^{14}C from the atmosphere. Thus, knowing the decay rate of ^{14}C, one may estimate the "conductivity" of the atmosphere-ocean interface for the transfer of CO_2. This conductivity (usually referred to as the "piston velocity") has dimensions of length/time. Conductivity multiplied by the atmospheric CO_2 concentration (moles/volume) gives the "invasion rate" (moles/area/time). Knowing the area of the oceans (3.6×10^{14} m^2), the global atmosphere to ocean rate is calculated to be 6.4 Pmoles/y. This parameter is not accurately determined. Bolin (1986) gives 18 ± 5 moles/m^2/y for the invasion rate, but even with such large uncertainties it is clear that the net accumulation rate of atmospheric CO_2 during the late twentieth century is only about

2–3% of the atmosphere-ocean exchange rate. As Degens (1989) points out, the fact that "the net flux of anthropogenic CO_2 into the oceans is only of the order of 2% of the natural exchange . . . explains some of the difficulties associated with the CO_2 problem."

The other major flow of carbon from the atmosphere is that driven by terrestrial photosynthesis. The dimensions of this process have been discussed in chapter 1. The value suggested there was 4.42 Pmoles/y for net primary production, giving an estimate of 7.73 Pmoles/y for gross primary production (the sum of NPP and the short term fixation of carbon that is then returned to the atmosphere by the "dark" respiratory activities of the autotrophs). As in the case of the atmosphere-ocean exchange, the value of this flow rate is not known with any accuracy. The reasons for the uncertainty are also discussed in the previous chapter. Nonetheless, it is clear that the terrestrial photosynthetic assimilation rate is between one and two orders of magnitude greater than the net accumulation rate indicated by atmospheric CO_2 measurements over the past thirty years. As in the case of the oceanic exchanges, we find that the natural process is much greater than the human impact.

It is important to keep in mind the quantitative relationships of the natural processes of ocean-atmosphere exchanges and terrestrial photosynthesis to the human input of CO_2 when considering the uncertainties concerning the accumulation of this gas in the atmosphere. Relatively small changes in the strengths of the major sinks and sources for atmospheric CO_2 could bring about proportionately large changes in the accumulation rate. It is thus not surprising that there is difficulty in determining what combination of changes in the input and output rates is responsible for anomalies in the CO_2 accumulation rate in the late 1980s (Conway et al. 1994; Francey et al. 1995; Keeling et al. 1995).

There is general agreement, however, that a major factor in the observed long-term increase in atmospheric CO_2 is the combustion of fossil carbon sources such as coal and petroleum hydrocarbons. Since the middle of the nineteenth century there has been an exponential increase in the use of fossil fuels. Until 1914 the semi-logarithmic plot of annual CO_2 emissions due to the combustion of fossil fuel against calendar year is linear, with a slope of 0.42/y. After 1914 there are irregularities in the slope that correspond to the major wars, to the economic depression of the 1930s, and, since 1979, to the major efforts in the industrialized nations to increase the efficiency of energy extraction from fossil fuels. The projection of this curve into the future depends

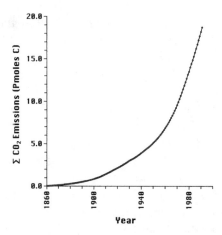

FIGURE 2.2 Global CO_2 emissions from fossil fuel consumption (1860–1949) and from fossil fuel consumption, cement manufacturing, and gas flaring (1950–1991). The summated emissions are obtained by adding the annual figures given for the earlier period in Keeling 1994 and, for the more recent period, in Marland et al. 1994.

heavily upon assumptions about the rate of economic growth in the developing countries.

In figure 2.2 I have presented the historical data of CO_2 emissions in integrated form. One advantage of this mode of presentation is that it facilitates an understanding of the consequences of the much-discussed policy options that would stabilize fossil fuel emissions at the level of any recent year (say, 1990). Notably, a stabilizing of emissions does not produce a stabilizing of atmospheric CO_2. Even if emissions level off, a major loading of CO_2 into the atmosphere would continue. The data of figure 2.2 suggest that the total of CO_2 emitted into the atmosphere by fossil fuel combustion through 1991 was 18.7 Pmoles of carbon. Bolin (1986) gives 15.25 ± 1.25 Pmoles of fossil fuel carbon emissions through 1984. This figure is one of the better determined components of the carbon budget. However, during the same time period covered by figure 2.2 there have been other significant changes in human activity with potential consequences for the biogeochemical cycle of carbon.

The most significant of these changes is deforestation. When forest is cleared for agricultural use, the carbon of the standing wood and of lit-

Dimensions of the Anthropogenic Perturbation   43

ter and roots is converted to CO_2. When deforestation is carried out by burning, however, significant amounts of carbon may be preserved as charcoal. After a forest is cleared, the level of soil carbon will fall by oxidation of humic materials (Harrison et al. 1993). It is important to distinguish between effects on biomass (and associated dead organic matter) and effects on productivity. When forest or grassland is converted to agricultural use, particularly when fertilizers are applied, the NPP of the land may increase considerably. However, the carbon so fixed is immediately made available to consumers (farm animals or directly for human consumption) so the new biome no longer constitutes a reservoir for carbon. On the other hand, when agricultural land reverts to forest, as has happened in some parts of North America, carbon is withdrawn from the atmosphere. The situation is more complex and much less well documented than the case of fossil fuel combustion. Extraction of coal and hydrocarbons and their subsequent use is subject to considerable bureaucratic supervision; the documentation thus generated provides a basis for geochemical quantitation. The clearing of forests and disposal of wood, on the one hand, and the abandonment of farmland with consequent vegetative succession, on the other, are not well documented. Thus the quantitative effects of these processes on the global carbon cycle are much more difficult to estimate (Mellilo et al. 1993; Dixon et al. 1994).

A further complication with respect to the role of the biota in the global carbon budget is the potential response of terrestrial plants to the increase of atmospheric CO_2. It is known that, under appropriate circumstances, plant growth can be stimulated by increased CO_2. The details of this "CO_2 fertilization" will be taken up in chapter 3, where the effect will be examined as an important potential feedback mechanism in global metabolism. The significance of the effect in modulating the recent changes (post-1850) is unclear (Gates 1985). An effect that stimulated the growth, primarily, of short-lived vegetation would have only marginal effects on the carbon balance (Botkin 1977; Lemon 1977; Oechel 1993) unless the extra growth contributes to the carbon pool of the soil. Similar considerations may apply to the global fertilization effects of increased mobilization of other nutrient elements, such as nitrogen and phosphorus.

Attempts have been made to quantify the overall effect of these various changes of deforestation, reforestation, and CO_2 fertilization on global biomass. In summarizing these attempts, Bolin (1986) suggests a net value of 12.5 Pmoles of carbon released into the atmosphere between 1860 and 1984, which is 82% of the anthropogenic carbon emissions owing to fossil fuel combustion. Bolin, however, qualifies this calculation of vegetation-related carbon changes with a rather high uncertainty: ± 4.2 Pmoles C. The sum of the two major inputs (fossil fuel combustion plus changed land use) is therefore 27.75 ± 5.45 Pmoles C. How much of this carbon has remained in the atmosphere? An answer to this question is an essential component of any credible prediction about the future course of CO_2 build-up in the atmosphere. Another major uncertainty arises in connection with this question. Although the progress of CO_2 accumulation between 1960 and the present day has been carefully monitored, there are no comparable data from previous years. Specifically, there are no accurate observations of atmospheric CO_2 in the mid nineteenth century. As each 1 ppmv CO_2 corresponds to 0.18 Pmoles CO_2, the estimate of the total CO_2 accumulated in the atmosphere since, say, 1860 depends upon the value assumed for atmospheric CO_2 at that date. If the value in 1860 was, for example, 260 ppmv, then the accumulation to 1984 would have been 14.8 Pmoles C. If the value in 1860 was 290 ppmv, then the corresponding accumulation would have been 9.4 Pmoles C. On the former assumption, the fraction of anthropogenic CO_2 remaining in the atmosphere is 53%; in the latter case, 34%. The only relevant observations are of the CO_2 content of the air bubbles in ice cores from the Greenland and Antarctic ice caps, which suggest a value in 1860 of 280 ± 5 ppmv (Neftel et al. 1985).

Another source of information concerning anthropogenic perturbation of the carbon cycle is the carbon isotope composition of tree rings. Tree-ring data provide a dated record of changes in the ^{13}C and ^{14}C content of the CO_2 from which the cellulose and lignins of the wood were synthesized. The least equivocal of these records is that of ^{14}C. This isotope is generated in the atmosphere by the interaction of cosmic ray protons with nitrogen; ^{14}C, like ^{12}C, is then incorporated into biomass by photosynthesis and, subsequently, into fossil fuels. ^{14}C has a half-life of 5700 years. Fossil fuels are many times older than that and, consequently, their ^{14}C content has decayed to negligible levels. CO_2 generated by combustion of fossil fuels contains no ^{14}C and therefore emissions of CO_2 from this source dilute the ^{14}C content of atmospheric CO_2. The wood synthesized from CO_2 reflects this dilution and the effect is clearly observable in dated tree rings from 1820 to 1950 (figure 2.3). Unfortunately, since 1954 the effect is obliterated by the injection of ^{14}C into the atmosphere through the testing of nuclear

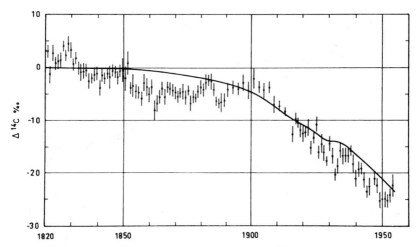

FIGURE 2.3 Dilution of the atmospheric content of ^{14}C by fossil fuel CO_2 between 1820 and 1954, as recorded in dated tree rings.
SOURCE: Figure 3.5 of Bolin 1986. Copyright © SCOPE 1986. Reprinted by permission of SCOPE.

weapons. Approximately 50% of the emissions from fossil fuels have taken place since 1950 and, had it not been for the effect of bomb testing, the ^{14}C dilution curve would by now have presented a much stronger signal.

It is similarly possible to infer the ^{13}C content of atmospheric CO_2 from measurements of this isotope in dated tree rings. Because the process of photosynthesis discriminates against the heavier isotope, all bioproducts—fossil fuels as well as wood—are depleted in ^{13}C relative to the CO_2 from which they were synthesized. When these bioproducts are oxidized to CO_2 either by combustion or respiration, the CO_2 returned to the atmosphere will be similarly depleted and will dilute the ^{13}C content of atmospheric CO_2. This dilution is observable in the ^{13}C content of tree rings (figure 2.4). Because this dilution arises from the return of isotopically light (depleted in ^{13}C) CO_2 to the atmosphere from both fossil fuel and from the organic carbon of the biota and soils, while the dilution of ^{14}C reflects only the emission of CO_2 from fossil fuel, it should be possible, in principle, to use the tree ring record of the two isotopes to deduce the relative contribution of these two major sources of CO_2 increase in the atmosphere. Such deconvolution of the record has been attempted (Stuiver 1978; Peng et al. 1983). The deduction is, however, far from straightforward. One important complication arises from the fact that the isotopic fractionation (discrimination against ^{13}C)

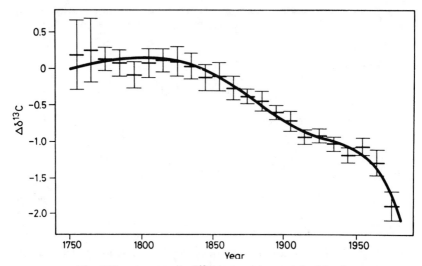

FIGURE 2.4 The 200-year record of ^{13}C in tree rings of the Northern Hemisphere. ^{13}C is diluted both by fossil fuel CO_2 and by biospheric CO_2.
SOURCE: Figure 9 of Freyer and Belacy 1983. Copyright © American Geophysical Union 1983. Reprinted by permission of the American Geophysical Union.

which provides the basis for the ^{13}C dilution effect is itself subject to variations in tree physiology and local environmental influences that cannot be allowed for.

Human impingement on the carbon cycle is thus evidenced in a number of ways.

1. Fossil fuel combustion has, since the mid nineteenth century, increased in exponential fashion (figure 2.2) and currently is injecting carbon into the atmosphere at a rate of about 0.49 Pmoles/y.
2. The CO_2 content of the atmosphere has increased from 315 ppmv in 1960 to 356 ppmv in 1992 (figure 2.1); the current rate of carbon increase is 0.25 Pmoles/y.
3. Between 1820 and 1950 the ^{14}C content of atmospheric CO_2 fell by about 2% (figure 2.3).
4. Between 1800 and 1960 the ^{13}C content of atmospheric CO_2 also fell by about 2% (figure 2.4).

These four sets of empirical data provide constraints to which any detailed model of the global carbon cycle must conform. Many such numerical models of the global carbon cycle have been developed. All such models agree that a significant transfer of anthropogenically produced CO_2 into the oceans must have occurred.

Uptake of CO_2 by the oceans depends upon two principal factors: the buffering capacity of surface waters and the rate at which surface water is transferred to the deeper parts of the ocean. For modeling purposes these processes can be calibrated against the observed oceanic depth profiles of pre-1950 ^{14}C, bomb-produced ^{3}H, and phosphate liberated from sedimenting organic detritus (Broecker and Peng 1982). Nonetheless, such models leave considerable uncertainty about the rate at which excess CO_2 will be disposed of by transfer into the oceans. The residual balance between the current rate of fossil fuel combustion (0.49 Pmoles C/y), plus the rate of CO_2 production attributable to changes in land use (0.13 Pmoles C/y) (Bolin 1986), minus the corresponding rate of accumulation of CO_2 in the atmosphere (0.25 Pmoles C/y) is 0.37 Pmoles C/y. It has generally been assumed that most of this carbon is taken up by the ocean.

The ocean model used by Peng et al. (1983) predicts an uptake of 0.29 Pmoles C/y. More recent calculations by Tans, Fung, and Takahashi (1990) are based on measurements of the air-sea pCO_2 differences and the small gradient of atmospheric CO_2 between the Northern and Southern hemispheres, which is attributable to the major inputs of industrial CO_2 in the north. These observations suggest a much lower value for oceanic uptake, 0.08 Pmoles C/y. Watson et al. (1991) and Smith et al. (1991) have pointed out that the method used by Tans and his colleagues fails to take into account the marked spatial and temporal variability in the air-sea CO_2 gradient, a variability that is brought about by phytoplankton blooms. Other corrections are needed to allow for riverine transport of carbon to the ocean (Sarmiento and Sundquist 1992) and for the existence of a nonanthropogenic north-south interhemispheric flow of carbon through the ocean (Broecker and Peng 1992). However, even after such corrections, the oceanic sink can account for less than 0.2 Pmoles of excess carbon uptake per year. There is thus a "missing sink" that must draw down excess carbon at about the same rate as does the oceanic sink. The models agree that this missing sink must be located in the Northern Hemisphere and that it is probably terrestrial (Ciais et al. 1995). The obvious candidate is the temperate and boreal forests, but carbon balance studies of these biomes do not support this suggestion (Houghton 1993). Until a finely structured map of the biologically driven air-sea gradient becomes available and other major uncertainties concerning the response of the global biota to increased CO_2 and to nutrient fertilization are resolved, further refinement of the oceanic and terrestrial models will probably not be decisive in closing the question of the global carbon budget. The current state of the problem is reviewed by Sundquist (1993) and by Siegenthaler and Sarmiento (1993).

In terms of the four-box scheme (figure 1.6) the human impact upon the carbon cycle is complex. Combustion of fossil fuels (which, as we have noted earlier, may be put within the box denoted "inorganic" or that labeled "bioproduct") enhances the process of mobilization or, if fossil fuel is thought of as bioproduct, regeneration. In both paths carbon is transferred to the nutrient compartment. The current rate of carbon mobilization owing to combustion of fossil fuels (0.49 Pmoles/y) is 40 to 50 times the estimated rate at which fossil organic carbon is oxidized in the unperturbed cycle. Clearance of forest lands also enhances the process of regeneration, but there are other effects. If wood is used for construction rather than burned, or if incomplete combustion leaves considerable amounts of charcoal, then some carbon moves from the biomass to the bioproduct compartment of the box model, rather than directly back to the nutrient box. If climax forest is replaced by rapidly growing shrubs and young trees, then the process of assimilation may be enhanced, which moves carbon from the nutrient box into biomass. Increases in atmospheric CO_2 and in other nutrients may enhance the rate of assimilation. From the point of view of the terrestrial biota, transfer of CO_2 into the sea is a mode of sequestration—that is, it is no longer available to terrestrial autotrophs. It seems probable that such transfer is quantitatively the most important route for the disposal of increased atmospheric CO_2. Unfortunately, the role of marine phytoplankton in disposing of this excess carbon is one of the least understood aspects of the problem. In most of the ocean, production is limited not by the availability of carbon but by the availability of phosphorus, nitrogen, or other nutrients such as iron (Martin et al. 1990; Martin et al. 1994) or zinc (Morel et al. 1994). It has thus been generally assumed that no marine biotic response would be expected to changes in dissolved inorganic carbon. This view has recently been challenged by Riebesell et al. (1993). In any event, if, as is the case for coastal waters, these nutrients are also on the rise, then increased assimilation in the sea, with subsequent increased sequestration by the sedimentary deposition of organic detritus, may be another significant change in the global carbon cycle.

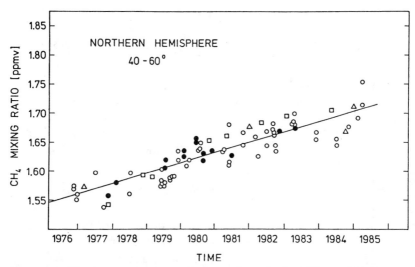

FIGURE 2.5 The increase in atmospheric methane in the 1970s and 1980s. Four different data sets are shown here, as represented by the different symbols for data points. All trend upward. For recent changes in this trend see Rudolph 1994.
SOURCE: Figure 4.1 of Bolle, Seiler, and Bolin 1986. Copyright © SCOPE 1986. Reprinted by permission of SCOPE.

Human Impact on the Carbon Cycle: Other Carbon Compounds of the Atmosphere

Carbon dioxide is not the only form of carbon in the atmosphere. Because CO_2 represents the most highly oxidized form of carbon, in Earth's oxygen-rich atmosphere, it would be expected to be the dominant gaseous form of carbon. And it is. But the atmosphere also contains carbon monoxide (CO) and methane (CH_4), albeit in much lower concentrations.

There are clear indications of human impact on the atmospheric level of CH_4. Consistent atmospheric observations are available only in the last two decades, but the upward trend is readily observable even over such a limited period of observations (figure 2.5). Longer term trends in atmospheric CH_4 can be deduced from analyses of air bubbles in ice cores from the Antarctic and Greenland ice caps (figure 2.6). Bolle, Seiler, and Bolin (1986) have pointed out the strong parallelism between the increase of atmospheric CH_4 and the growth of the human population—clear evidence that the atmospheric increase is a response to anthro-

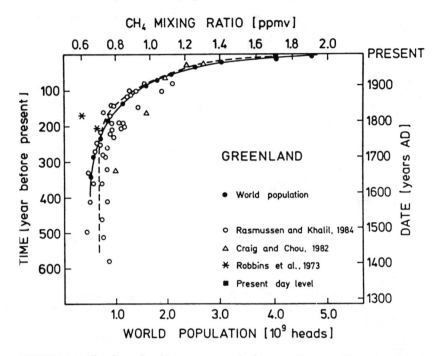

FIGURE 2.6 The Greenland ice core record of atmospheric methane in relation to world human population.
SOURCE: Figure 4.3 of Bolle, Seiler, and Bolin 1986. Copyright © SCOPE 1986. Reprinted by permission of SCOPE.

pogenic pressures. However in the late 1980s and early 1990s the rate of increase of atmospheric CH_4 has fallen markedly (Rudolph 1994). The reasons for this slowing are unclear, reflecting the marked uncertainties concerning the budget of atmospheric CH_4.

The principal sources of methane are biological. CH_4 is produced by microorganisms in anaerobic, reducing environments. Environments that have been implicated as significant global sources of CH_4 are the rumens of herbivores, where microbial breakdown of cellulose produces large amounts of acetate, a substrate for methanogenesis. Similarly, digestion of cellulose by microbial symbionts provides the energy supply for termites, and in this case also CH_4 is a by-product. Waterlogged, anaerobic soils—such as rice paddies and natural wetlands—are also important sources of biologically produced methane. Finally, municipal solid waste sites and biogas digesters offer rich, anoxic environments for CH_4 production.

Methane also occurs in association with fossil carbon and hydrocarbon deposits. It is released into the atmosphere as a by-product of coal mining and natural gas extraction and distribution. Another significant input of CH_4 into the atmosphere occurs as a result of the burning of biomass (Crutzen and Andreae 1990). Quantitation of these inputs is difficult. The inputs are calculated as the product of the estimated extent of the source multiplied by its estimated intensity. In all cases these estimates are attended by considerable uncertainty. The values in the literature are reviewed by Bolle, Seiler, and Bolin (1986), who also make an attempt to estimate the change in these inputs from 1940 to 1980 (see also the chart in Rudolph 1994). It has recently proved possible to measure the ^{13}C and ^{14}C content of atmospheric CH_4 (Craig et al. 1988; Wahlen et al. 1989). Such isotopic measurements are of potential use, as explained in the previous section, for disentangling the relative contribution of fossil and biological sources.

Methane is removed from the atmosphere primarily by reaction with the photochemically produced radical OH to yield the methyl radical CH_3, which is converted in a series of reactions to carbon monoxide, CO. The reaction pathway depends on the tropospheric concentration of nitric oxide, and it may be important in the generation of tropospheric ozone (which will be discussed later in this chapter). Because oxidation by OH is the quantitatively dominant sink for CH_4, the recent reevaluation (Vaghjiani and Ravishankara 1991) of the rate constant for the oxidation of CH_4 by OH will bring about corresponding changes in estimates of the turnover time of the atmospheric methane pool.

The oxidation of CH_4 by OH constitutes a major source of atmospheric CO. In the last 150 years the incomplete combustion of fossil fuels must have added a significant anthropogenic input of CO, especially from the exhausts of motor vehicles (Freyer 1979). The third major source of CO is located in the tropics. This tropical source of CO is derived in part from the photochemical oxidation of hydrocarbons (isoprene and terpene) produced by trees, and in part from another human activity—namely, the burning of biomass (Crutzen 1983; Newell et al. 1989). As in the case of CH_4, the principal removal process for CO is reaction with OH. In this case also, the reaction pathway and its effects on OH and O_3 levels will be determined by the level of NO. Estimates of the input rates and removal rates are given in table 2.1. In the early 1990s there has been a marked decrease in atmospheric CO, though whether this drop is due to decreased sources or increased sinks or some combination of both remains obscure (Novelli et al. 1994). Khalli and Rasmussen (1994) concluded, "What is clear is that the rather sud-

TABLE 2.1
Inputs and Outputs for Atmospheric Carbon Monoxide

	Tmoles/y
Combustion of biomass	14–57
Industrial combustion	23
Plant metabolism	0.7–7
Tropospheric oxidation of CH_4	13–33
Tropospheric oxidation of other hydrocarbons	14–46
TOTAL SOURCE STRENGTH	65–166
Uptake by soils	16
Tropospheric oxidation by OH	107
TOTAL SINK STRENGTH	123

den and rapid decline in CO is an indicator of global change with far-reaching implications regarding the effect of human activities on the global environment."

Human Impact on the Nitrogen Cycle

Anthropogenic effects are also clearly observable in the global cycle of nitrogen. A major difference between the cycles of carbon and nitrogen lies in the nature and location of the inorganic (or abiotic) reservoir. For carbon, the inorganic compartment of the biogeochemical cycle consists largely of hydrated forms of CO_2 (HCO_3^- and CO_3^{2-}) dissolved in the ocean. In contrast, the predominant inorganic pool of nitrogen is dinitrogen, N_2, found primarily in the atmosphere, where it is by far the most abundant gas. The amount of N_2 dissolved in the oceans is only about 0.5% of the size of the atmospheric pool.

Dinitrogen is a relatively unreactive compound. It is this lack of reactivity that accounts for the fact that, despite its geochemical abundance, nitrogen is often the limiting factor in the processes that are aggregated under the heading "assimilation" in the four-box scheme of figure 1.6. Attention is therefore focused on those processes whereby N_2 is converted to biologically available molecular species that make up the nutrient pool in this cycle, namely, ammonia, NH_3, and nitrate (NO_3^-).

The naturally occurring processes of reduction of N_2 to NH_3 and of its oxidation to NO_3^- were discussed in chapter 1. Because both pathways chemically transform inorganic nitrogen into nutrients accessible to life, these processes are denoted as "mobilization" in the global nitrogen cycle. Both reduction and oxidation of N_2 are markedly augmented by human activities. The increase in reduction to NH_3 is brought about deliberately in order to increase agricultural productivity; the oxidation to nitrogen oxides (NO_X) occurs inadvertently, as a by-product of combustion. The rate of industrial production of NH_3 is known more precisely than that of the biological process. For 1979/80 the U.N. Statistical Yearbook gives 4.3 Tatoms of nitrogen per year. Delwiche and Likens (1977) give 2.9 Tatoms N/y. Both figures are comparable with the lowest estimates of global biological fixation of nitrogen (3 Tatoms/y) and between 20% and 30% of the highest estimates (14 Tatoms/y) (Rosswall 1983). It is important to note that there is also a deliberate attempt to increase the biological process of N_2 fixation by changing agricultural practices, such as the increased planting of legumes. If genetic engineering were to bring about a much wider distribution of the genes for N_2 fixation, this might markedly change the rate of biological fixation.

Human activity increases the production of NO_X by accelerating the thermodynamically favored combination of atmospheric O_2 and N_2 at the high temperatures associated with all combustion processes. The magnitude of the global summation of all these diverse processes is not well known. Estimates vary from 0.7 to 2.9 Tatoms N/y for oxidation related to industrial combustion and fossil fuel burning (Rosswall 1983; Simpson 1977). But NO_X production is also associated with the burning of plant material, some of which would be natural fire and some of which would be associated with human activities such as forest clearing. Estimates for the strength of this source range from 0.7 to 14 Tatoms N/y (Rosswall 1983). Nitric oxide (NO) and nitrogen dioxide (NO_2) are removed from the atmosphere in a few days (Crutzen 1983) by reactions involving ozone and the hydroxyl radical

$$NO + O_3 \rightarrow NO_2$$
$$NO_2 + OH \rightarrow HNO_3$$

The end-product, nitric acid, is highly water soluble and is washed out of the atmosphere by rain. In contrast, nitrous oxide, N_2O, has a much longer life time in the atmosphere, 100–200 years, and is present in the atmosphere at much higher concentrations (300 ppbv; 10,000 × the concentration of NO and NO_2) (Bolle et al. 1986). N_2O is produced in the same high temperature conditions in which the other nitrogen oxides are formed, but the processes of biological nitrification and denitrification (figure 1.12) play a much more quantitatively important role in the production of this atmospheric component.

After N_2, N_2O is the most abundant form of nitrogen in the atmosphere. There has been much discussion of the sinks and sources of N_2O and the ways in which they are being altered by human activity. Unlike the trace gases of the carbon cycle (CH_4 and CO) and the other nitrogen oxides, N_2O does not react directly with OH or O_3. No significant tropospheric sink for N_2O has been identified (Bolle et al. 1986). N_2O is destroyed in the stratosphere, either by photolysis to N_2 and O or by oxidation by atomic oxygen in an excited state, the latter reaction yielding NO among the products. The oxidation of N_2O is believed to constitute the principal source of stratospheric NO and NO_2. The sources of N_2O are not well quantified. As shown in figure 1.12, both nitrification (the oxidation of NH_3 to NO_3^-) and denitrification (the reduction of NO_3^- to NH_3) produce N_2O as a by-product. The former, oxidative process may be dominant in agricultural lands treated with ammonia fertilizers (Conrad et al. 1983). Whichever biochemical route is dominant, it seems clear that N_2O production rates are being significantly enhanced by human activities.

Bolle, Seiler, and Bolin (1986) give their best estimate of global emissions of N_2O as containing 0.86 to 1.07 Tatoms of nitrogen per year. Of this figure, 0.06 to 0.21 Tatoms might arise from the conversion of undisturbed land to agricultural use and from the application of fertilizer nitrogen. These effects would account for between 5 and 24% of the input. Measurements from the mid 1970s to the mid 1980s showed an annual increase of about 0.21 to 0.32 Tatoms of nitrogen in the form of N_2O in the atmosphere. As the atmosphere contains a total of about 53 Tmoles of N_2O, the annual increase is about 0.2 to 0.3%. Such an increase is not out of line with the upper estimates of the anthropogenic inputs. N_2O production rates provide only a lower bound for the rate of sequestration of nitrogen—that is, conversion into a form not directly assimilable by life—through the pathways of nitrification and denitrification. The other major product of the denitrification process is N_2, which until recently has not been accounted for. Fluxes of N_2 are now being reported for both marine (Devol 1991) and terrestrial (Kuhlbusch et al. 1991) environments. These measurements will be important in achieving a direct estimate of global denitrification. The proportions of N_2 and N_2O produced in denitrification are highly variable. Estimates of the total loss of NO_3^- from land and from the ocean are highly scattered. For example, Rosswall (1983) reported 3.1–27.9 Tatoms N/y for the former and 0–23.6 Tatoms for the latter. It would therefore be impossible to detect any anthropogenic perturbation against such an imprecisely quantified background of natural processes. That such perturbation is occurring, nevertheless, is evidenced by the increase in atmospheric N_2O, but that figure cannot be used to quantify human impact on the rate of sequestration in the global cycle of nitrogen.

Crutzen and Andreae (1990) have recently reviewed their findings concerning loss of fixed nitrogen brought about by the burning of biomass in the tropics. Estimates of such pyrodenitrification, of which the major product is N_2, are as high as 1 Tatom N/y (Kuhlbusch et al. 1991). If correct, this could signal an important loss of nutrient N in the tropics and a significant proportion of the global flux. It should be noted however that N_2O is only a minor constituent of the gaseous compounds of N produced in such fires. Overall, biomass burning is considered to be a relatively minor global source of N_2O (Cofer et al. 1991).

Human Impact on the Phosphorus Cycle

In the case of the global cycle of phosphorus, the principal human impact is upon the rate of mobilization—that is, upon the rate at which an element is made available to autotrophs by transformation and translocation. Rocks rich in phosphate—mostly apatite, $Ca_5F(PO_4)_3$—form in freshwater or marine environments. Once phosphate is deposited as sediment and lithified into rock, no organisms other than humans are capable of extracting the nutrient from the deposits. The rate of return of nutrients to the biosphere occurs by physical and chemical action that releases particles through erosion and weathering. That constraint has, of course, changed with the impact of human activity.

Phosphates are currently being mined, largely for use as fertilizer, at a rate of 0.45 Tmoles P/y (Richey 1983). This rate is much higher than the undisturbed rate at which phosphorus flows through the sedimentary cycle (see table 1.1). Even if we use a lower estimate of 100 My for the turnover time of sedimented phosphorus (Broecker and Peng 1982), the current rate of mining is about 1.7 × the mobilization rate of the geochemical cycle. On the other hand, comparison with the figures for rates in the undisturbed cycle given in Richey 1983 show that this rate is only a small percentage of the estimated global rate for transfers between the terrestrial biota and soils. The two most important effects of this increase in mobilization are the gains in agricultural productivity—which is, of course, the reason why the phosphate rock is being mined—and, second, the inadvertent eutrophication of lakes and coastal waters that occurs because much of the added phosphate is leached away from its site of application. In terms of the four-box scheme (figure 1.6), the effect of human activity has been to increase the size of the nutrient pool. This, in turn, as predicted by Liebig's Propositions 40 and 41 (to be discussed in chapter 6), leads to gains in primary production (assimilation), both desirable—increased agricultural yield—and undesirable—increased plant, algal, or bacterial growth in natural waters.

The question arises with respect to the increased rates of mobilization of both phosphorus (by mining) and of nitrogen (by industrial production of NH_3 and by the production of NO_X in combustion processes) as to whether these increases could bring about some measure of eutrophication with an impact at the global scale. Is the human-caused mobilization of nitrogen and phosphorus out of the inorganic pool and into the nutrient pool bringing about an increase in global biomass? To the extent that global primary productivity is limited by N or P or both, an increase in the availability of these nutrients might indeed increase the production of biomass. Thus, either as a result of mass action effects (to be discussed in chapter 6) or through stoichiometric constraints (Redfield ratios, also in chapter 6), perturbation of the global cycles of N or P might indirectly affect the carbon cycle.

Specifically, by increasing global production of biomass, fertilization by N or P might strengthen the biological sink for CO_2 and thus ameliorate its build-up in the atmosphere (and solve the problem of the "missing sink," discussed earlier). Since this scenario was suggested by Garrels, Mackenzie, and Hunt in 1975, a number of authors have explored this possibility (Simpson 1977; Broecker et al. 1979; Peterson and Mellilo 1985; Schindler and Bayley 1993; Hudson et al. 1994). Bolin (1986) lists two major uncertainties in projections of future atmospheric CO_2 levels: "inadequate knowledge about . . . the sensitivity of marine primary productivity to changes of nutrient availability in surface waters" and "fertilization and increase of biomass and organic matter in soils in terrestrial ecosystems due to . . . deposition of nutrients emitted from anthropogenic sources."

The obvious difficulties in ascertaining whether human-induced mobilization of N and P are measurably offsetting human-induced injection of carbon into the atmosphere arise from uncertainties about which biomes are capable of response to nutrient deposition. Also crucial are quantitative estimates of the extent to which the input of nutrients to such biomes is actually being augmented by human action, and the ratio of assimilation (carbon/nutrient) in any increment of biomass. Among the papers just cited, estimates of anthropogenic loading of nitrogen to northern temperate and boreal forests vary from 0.4 to 1.5 Tatoms N/y. Estimates of C/N ratios of the increased biomass range from 17 to 100. More generally, the extent of any such nutrient effect will depend upon where the system is on the curve relating growth to nutrient concentration (to be discussed in chapter 6) and how close the system is to any other upper limit for growth. The four-box scheme (figure 1.6) can be used quantitatively to confirm the intuitive expectation that if the rate of assimilation is close to saturation with respect to nutrient, or if the biomass pool is near some limit set by other environmental variables, then there is no potential for increased productivity to generate an additional sink for anthropogenic CO_2. Conversely, when neither of these constraints apply, fertilization can have a marked effect on the atmospheric content of CO_2.

Human Impact on the Sulfur Cycle

The last global elemental cycle to be considered in this chapter is that of sulfur. As in the case of phosphorus, the principal effect of human activity is upon mobilization of sulfur from the lithosphere—placed in the inorganic compartment of our four-box scheme. Much of this mobilization is inadvertent. Some minerals that contain sulfur are deliberately mined, primarily for agricultural purposes. Gypsum ($CaSO_4 \cdot 2H_2O$) is an example. But the amount of sulfur incidentally liberated in association with the smelting of sulfide ores (such as, chalcopyrite, $CuFeS_2$) and in association with the extraction and combustion of fossil fuels that also contain sulfur exceeds the quantities liberated expressly for obtaining sulfur. A further mobilization of sulfur out of the lithosphere must be occurring as a consequence of the anthropogenic acceleration of weathering (table 2.2).

Clearly, the anthropogenic sources of sulfur emitted to the hydrosphere and atmosphere are comparable in strength to the natural sources. Unlike that of phosphorus, but similarly to nitrogen, the biogeochemical cycle of sulfur includes important atmospheric components. The three most quantitatively important are carbonyl sulfide, COS (500 pptv), sulfur dioxide, SO_2 (10–200 pptv), and dimethyl sulfide, $(CH_3)_2S$, (0–100 pptv). But other gaseous forms of sulfur—such as hydrogen sulfide, H_2S, and carbon disulfide, CS_2—are also atmospheric constituents. The end product of the oxidation of these compounds is sulfate (SO_4^{2-}) or methyl substituted sulfates ($CH_3SO_3^-$). Sulfate is not usually a limiting factor in the growth of primary producers—whether terrestrial, fresh-

TABLE 2.2
Emission of Sulfur to the Atmosphere and Hydrosphere

Source	Rate (Tmoles/y)	Product
Weathering	1.88	SO_4^{2-}
Volcanoes	0.14	SO_2
Biogenic: oceans	1.25	$(CH_3)_2.S$
Biogenic: land	1.25	$(CH_3)_2.S; H_2S$
Total natural	4.52	
Raw materials	1.62	SO_4^{2-}
Increase in weathering	0.94	SO_4^{2-}
Combustion	1.87	SO_2
Smelting	0.31	SO_2
Total anthropogenic	4.74	

water, or marine—so the mobilization of sulfur by human activity does not bring about eutrophication. However, in the absence of any countervailing cation, all of the anthropogenic input of SO_2 may be expected to end up as H_2SO_4 (sulfuric acid). Sulfuric acid acts as an atmospheric source of protons (H^+) and, thus, of acid rain.

Overall, for the biogeochemical cycles of C, N, P, and S there is evidence that the rates of processes subserving the cycles and the size of some of the constituent compartments are indeed being affected by human activity. Table 2.3 summarizes the anthropogenic effects on mobilization in these four cycles. A number of consequences of these changes give cause for concern. Eutrophication at the local level, as evidenced by increased plant growth in lakes and coastal waters, can markedly affect the species distribution in such ecosystems. Impacts on the levels of atmospheric constituents have raised widespread concern because of three indirect effects: (1) changes in the pH of precipitation; (2) changes in the O_3 content of the atmosphere; (3) changes in the radiative balance of the atmosphere.

Acid Rain

It could be argued that the problem of "acid rain" lies somewhat outside the scope of this monograph because the acidification of terrestrial waters is a regional rather than a global phenomenon. However, many globally distributed biomes are being affected by acid rain. The gases involved in acidification are the oxides of nitrogen and sulfur. These gases have short half-lives (1.5 days for NO_x, 5 days for SO_2). As a consequence, their concentrations fall as a function of distance from their

TABLE 2.3
Mobilization Rates in Global Biogeochemical Cycles of Carbon, Nitrogen, Phosphorus, and Sulfur

Element	Steady-State Rate (Tmoles/y)	Human Perturbation (Tmoles/y)
C	10	430
N	3–14	4–21
P	0.065	0.45
S	2	4.7

sources. After atmospheric transport over 1500 Km (E ↔ W), the concentration of NO_X is reduced by 70%. The corresponding transfer distance for SO_2 is 5000 Km (Crutzen 1983: table 3.1). The end products of atmospheric oxidation of NO_X and SO_2 are nitric acid (HNO_3) and sulfuric acid (H_2SO_4). These are strong acids and constitute the major external inputs of protons to ecosystems in northeastern North America.

In northwestern Europe the major route of deposition of sulfur and nitrogen is not as sulfuric acid and nitric acid dissolved in rain water, but rather as gases and aerosols. However, hydration and oxidation processes after deposition produce the same acids, and the effects on the proton budget of affected ecosystems on both continents are similar (Van Breemen et al. 1984). Not only the input is regional, depending on proximity to sources, the effects of such proton input are also regional, depending upon the capacity of the water and soil in a given ecosystem to neutralize acids.

From the viewpoint of global metabolism it is important to note that there are major biogenic sources that lead to the production of both HNO_3 and H_2SO_4. Nitrification brings about a net formation of protons:

$$2NH_4^+ + 4O_2 \rightarrow 4H^+ + 2NO_3^- + 2H_2O$$

Consequently, any ecosystem in which reduced forms of N, from either external or internal sources, are being oxidized to NO_3^- will generate protons (Gorham et al. 1979). Such effects can be very marked following deforestation, when the exposed forest floor is decomposing (Vitousek 1983). The external input of protons is therefore not the only way in which human activity can impinge upon the acid/base balance of ecosystems.

Only in recent years has it been realized that there are significant biogenic sources of reduced compounds of sulfur that are oxidized in the atmosphere to SO_2 and H_2SO_4. The significance of biogenic production of $(CH_3)_2S$ (dimethyl sulfide, DMS) in global metabolism will be discussed in chapter 3. For now, it is important to recognize that DMS production in coastal waters may contribute appreciably to the input of SO_2 and H_2SO_4 to the adjacent land masses. However, this biogenic source of acidic precipitation is not itself free from anthropogenic perturbation. Lovelock (1988) has suggested two ways in which human activity may be increasing the input of acid from this biogenic source. First, the rate of DMS oxidation may be increased in polluted air; therefore the SO_2 generated is concentrated in a more limited area adjacent to the source of DMS rather than being transported over longer distances and diluted over a wider area. Second, the rate of DMS production in coastal waters may be increased as coastal algal blooms form as a consequence of riverine transport of nutrients to the continental shelf.

Ozone

The second cause for concern about the anthropogenic perturbation of the chemistry of the atmosphere arises from indirect effects on atmospheric ozone (Crutzen 1983; Bolle et al. 1986). Ozone, O_3, is not uniformly distributed in the atmosphere; about 90% is in the stratosphere (>15 Km), where the mixing ratio may reach 1 ppmv. In the lower atmosphere (troposphere) the level of ozone in uncontaminated air is much lower. A representative value for clean air is 30 ppbv (Hough and Derwent 1990), but there are considerable spatial variations related to the effects of anthropogenic emissions.

Ozone is produced by the reaction of atomic oxygen, O, with dioxygen, O_2. In the stratosphere, the primary source of O is the photolysis of O_2 by solar ultraviolet radiation (<310 nm). The radiation that is responsible for the photolysis of O_2 does not penetrate into the troposphere. In clean tropospheric air there are two sources of ozone: influx from the stratosphere and reaction of O_2 with O produced by the photolysis of NO_2, the latter being brought by longer wavelength solar radiation (<400 nm).

As we have seen earlier in this chapter, human impact on the biogeochemical cycle of nitrogen can markedly increase the input of nitrogen oxides into the atmosphere. Consequently, large increases of tropospheric O_3 can be measured downwind from industrial centers, particularly in summer. The direct production of NO_X by high temperature

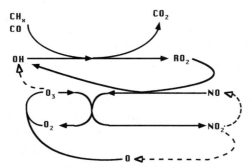

FIGURE 2.7 Reaction scheme for the tropospheric oxidation of CO and hydrocarbons. The dotted lines represent light-driven reactions, two of which produce the highly reactive radicals O and OH. R represents an acyl group. Note the role of nitrogen oxides in these cycles.

sources, such as the internal combustion engines of motor vehicles, is a major factor in this increase. The second product of photolysis of NO_2 is NO, which is oxidized back to NO_2 by reaction with O_3. A complex set of photochemically driven catalytic cycles connects the NO_X level to the atmospheric oxidation of CH_4 (and other hydrocarbons) and CO. These oxidations are initiated by attack by the hydroxyl radical, OH, which is replenished as a consequence of the photolysis of O_3. Because the hydroxyl radical is regenerated in the reaction sequences themselves, it plays a catalytic role. Peroxyl (HO_2) and substituted peroxyl radicals (RO_2) are intermediates in the reaction sequences. When the concentration of NO is sufficiently high, it can compete for RO_2, and a new catalytic cycle is generated that brings about a net production of O_3. The alternative cycles for the oxidation of CO are depicted in figure 2.7. One of the principal sources of atmospheric CO is the oxidation of CH_4 by similar cycles.

The conditions for the production of excess tropospheric O_3 are, therefore, input of NO_X, input of CO and hydrocarbons, and sunlight. Other products formed under these conditions are peroxyacyl nitrates, which are strong oxidizing agents. There is considerable concern that peroxyacyl nitrates, as well as ozone itself, may be acting synergistically with acid precipitation to injure vegetation downwind from sources of pollution (Hewitt et al. 1990).

High levels of ozone near the ground can harm the biota because of its high chemical reactivity. However, the 90% of atmospheric ozone that is in the stratosphere has an important physical property which is beneficial for the biota. Ozone has a strong absorption band in the ultraviolet region, which acts as a protective filter against radiation that is potentially mutagenic. For some years there has been concern that human activities could act to decrease the level of stratospheric ozone and thereby increase the penetration of harmful radiation to ground level. These fears now appear to have been confirmed (Kerr and McElroy 1993). In the stratosphere, as in the troposphere, O_3 levels are controlled primarily by the NO_X levels. The primary source of stratospheric NO_X is the relatively stable N_2O. Although there has been some concern expressed that the increase in N_2O associated with changed agricultural practices could bring about a decrease in stratospheric O_3, the current opinion is that this increase does not constitute a significant threat to the O_3 level (Crutzen 1983). Greater concern attaches to the anthropogenic input of persistent chlorinated (and brominated) compounds that are very slowly destroyed in the stratosphere to yield Cl and ClO, which can participate in catalytic cycles that lead to the net loss of O_3. There is now widespread agreement that the potential threat is so great that the production of these halogenated gases should be phased out.

Ozone levels are variable in time and space, so it is difficult to establish that there has been any decrease in mean global stratospheric O_3. It is, however, abundantly confirmed that O_3 losses can be particularly marked in the Antarctic at the end of winter, leading to the much discussed "ozone hole." The intensity of the hole over the South Pole has increased steadily since the mid 1970s (Solomon 1998, 1990). The mechanism of these losses is complex. Nevertheless, it is established that ozone loss is intensified by the input of compounds containing Cl and Br (Proffitt et al. 1989). NO_X slows the Cl and ClO catalyzed cycles of O_3 destruction. It appears that in the winter NO_X is removed as HNO_3 by dissolution in aerosols. Thus, when sunlight returns, the photochemical reactions that produce Cl and ClO are unrestrained until stratospheric warming evaporates the aerosols (Crutzen and Arnold 1986).

The compounds responsible for such dramatic effects on stratospheric O_3 are not natural products of the global biota (Anderson et al. 1991). Thus they may provide the first clear instance of novel products of human activity impinging upon global biogeochemical cycles (Häder et al. 1995; Zepp et al. 1995).

TABLE 2.4
Relative Effects of 1990 Emissions Integrated Over a 100-Year Time Horizon

Greenhouse Gas	Temperature Effect (% of total)
CO_2	61
CH_4	15
CFCs	11
N_2O and NO_X	10
Others	3

SOURCE: Shine et al. 1990.

The Greenhouse Effect

The third cause for concern about human impacts on atmospheric chemistry relates to the role of some anthropogenic emissions in determining the thermal balance of Earth. Solar radiation reaching Earth has a different spectral distribution than the radiation emitted by Earth. As a much cooler body than the sun, Earth radiates primarily in the infrared. Much of this radiation is absorbed by atmospheric constituents, particularly H_2O and CO_2. This trapping of energy, so that not all of the energy emitted by the surface escapes to space, is known as the greenhouse effect. Satellites have made possible direct observation of the total outward radiative flux at the top of the atmosphere (Raval and Ramanathan 1989). Comparison of the measured flux with the calculated black body radiation of the surface shows that about 42% of the emitted radiation is trapped—35% by atmospheric gases, 7% by clouds. Increases in the concentration of any of the so-called greenhouse gases could bring about an increase in the global mean temperature, with associated changes in climate.

Lashof and Ahuja (1990) have pointed out that the contribution of a given gas to global warming can be expressed as a function of three values: its emission strength, its ability to absorb infrared radiation, and its atmospheric residence time. Although CO_2 is a weaker absorber of infrared than are other anthropogenically affected gases (notably, methane), its high emission strength and long residence time combine to make CO_2 the most significant potential modifier of global temperature. Table 2.4 shows the relative thermal impacts of various anthropogenic emissions into the atmosphere. Of these, CO_2 would account for 61% of any temperature increase over the next hundred years if anthropogenic emissions continue at 1990 levels.

The greatest cause for uncertainty concerning changes in global temperature in the next century lies in the area of socio-economics (Wigley et al. 1996). Will CO_2 emissions be stabilized? Recall that even stabilization at 1990 levels implies a continuing anthropogenic input of carbon of about 0.5 Pmoles/y into the atmosphere. Or will CO_2 emissions increase as major nations such as China and India seek to achieve higher levels of industrialization with concomitantly high usage of fossil fuels.

There is little agreement about the magnitude of the expected temperature increase, even if rates of emissions were stabilized at 1990 levels. Estimates have ranged from a few tenths of a degree to more than 5°C. The major problem concerns the effect of clouds on the global radiation budget (Ramanathan et al. 1989; Cess et al. 1989; Charlson and Wigley 1994). Clouds not only trap outgoing infrared radiation; they also reflect incoming solar radiation. The net effect depends on such factors as the proportion of ice-dominated versus water-dominated clouds (Mitchell et al. 1989), droplet size, and height above the surface (Slingo 1990). The significance of clouds in the regulation of global temperature will be revisited in the next chapter.

To sum up: perturbation of the global biogeochemical cycles is being brought about by a greatly increased human population engaged in new technologies and in massive exploitation of Earth's resources. On a global scale the most obvious effects are on the chemistry of minor constituents of the atmosphere that represent small pools with high turnover rates. The importance of these changes lies in their side effects on the pH of precipitation, the transparency of the atmosphere to ultraviolet radiation, and the thermal balance of the atmosphere (greenhouse effect). It is the small pool sizes of these minor constituents that renders them vulnerable to anthropogenic perturbation. Similarly, because the size of the reservoirs is small, these minor constituents of the atmosphere may have been susceptible to nonanthropogenic impacts during the history of the planet. Nonetheless, as we saw in chapter 1, the planet appears to have remained habitable over much of geological history. What are the factors that have maintained habitability? Are the controls geochemical? Or are the global biota involved in keeping Earth habitable? The following chapters will consider a variety of answers that have been given to such questions.

3

The Life Boundary and Environmental Homeostasis

Chapter 1 reviewed the evidence suggesting that what might be termed *the global metabolic system* is from time to time subjected to three kinds of perturbing forces. These arise from (1) collisional events, (2) periodicities in the input of solar energy, and (3) tectonic cycling. The geological record shows that these impingements have had major effects on the planetary biogeochemical cycles. Nonetheless, the geological record also shows a persistence of biochemical activity over a period extending back to 3.8 Gya. In this chapter we turn to this "curious aspect of Earth history," as it is referred to by Holland (1984). In trying to understand the habitability of Earth, the longevity and continuity of life on this planet is a key issue for empirical investigation and theoretical explanation.

The primary geochemical evidence is given by the ^{13}C content of sedimentary organic carbon (Schidlowski et al. 1983). The autotrophic fixation of CO_2 into organic carbon, primarily brought about by ribulose-1,5-bisphosphate carboxylase, discriminates against the heavier isotope of carbon (^{13}C) in favor of the abundant isotope (^{12}C). The resulting biomass is thus depleted in ^{13}C by about $-25\%_{00}$. Consequently, the presence of carbon depleted in ^{13}C is an indication of autotrophic activity. No known inorganic process brings about isotopic fractionation of this magnitude. The difference between the ^{13}C content of organic carbon and the CO_2 from which it is formed is thus diagnostic of autotrophic assimilation.

66 The Life Boundary and Environmental Homeostasis

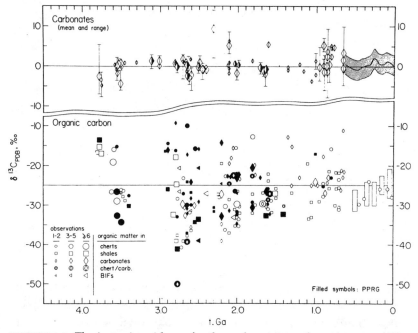

FIGURE 3.1 The isotopic evidence for the early origin and persistence of life in Earth's history. The decreased ^{13}C content of organic carbon compared to contemporaneous carbonates is indicative of autotrophic biological fixation of CO_2 and extends in an unbroken record from the earliest samples (3.8 Gya).
SOURCE: Figure 7.3 of Schidlowski, Hayes, and Kaplan 1983. Copyright © Princeton University Press 1983. Reprinted by permission of Princeton University Press.

Figure 3.1 summarizes the geological findings of isotopic fractionation that imply the unbroken history of life. It is noteworthy that the oldest preserved organic carbon has a ^{13}C content (compared against same-age inorganic carbonates) consistent with a biological origin. Figure 3.1 also shows that the sedimentary carbonates, which reflect the isotopic composition of the marine inorganic carbon from which they were precipitated, have changed their isotope content very little during the whole period. The changes in ^{13}C during the Phanerozoic (the past 600 my), shown in figure 3.1, are not large enough to affect the general argument. In any case, as we discussed in chapter 1, these changes are themselves attributable to the same process of biological fractionation. Although it is the difference between the ^{13}C contents of carbonates and presumed organic carbon that signifies the autotrophic origin of the latter, it is possible to insert the absolute values (about 0‰ for carbonates, −27‰ for organic carbon, and −5‰ for the primordial carbon of the mantle) into a mass balance equation. We can thus calculate that, from the earliest known appearance of sedimentary organic carbon, the ratio of organic carbon to carbonate being deposited in sedimentary rocks has remained constant at about 0.25 (Holser et al. 1988).

There is thus geochemical evidence reaching back 3.5 to 3.8 Gy for the existence of a biogeochemical cycle of carbon that was fundamentally similar to that operating on Earth today. This geochemical evidence, taken together with the morphological evidence of the existence of stromatolites and microfossils extending back to 3.5 Gya, implies that a habitable global environment has existed continuously over all of this period. At the Precambrian-Cambrian (Proterozoic-Phanerozoic) boundary, about 600 Mya, the first metazoan fauna appear in the fossil record. If one assumes an aerobic physiology for these organisms similar to that of modern small invertebrates (indeed, of all members of the animal kingdom), then the atmosphere at that time must have contained O_2 at a partial pressure of at least 0.02 atm (10% of the present level).

Despite the mass extinction events during the Phanerozoic era, the fossil record of complex metazoans is unbroken, and O_2 levels can thus never have fallen below levels adequate to support such a fauna. The persistence of a flourishing land-based metazoan fauna since Devonian times (350 Mya) suggests that atmospheric levels of oxygen have been relatively constant, not far from today's value. We have seen also in chapter 1, and it will be reiterated later in this chapter, that the oxygen of the atmosphere rises in stoichiometric equivalence to the deposition of reduced products of photosynthesis in marine and other sediments. The relatively constant ratio between the deposition of inorganic and organic carbon (suggested by their ^{13}C contents) during the Phanerozoic does not indicate any marked gains or losses of O_2 due to the operation of the carbon cycle. Any such fluctuations may have been partially compensated for by redox shifts in the sulfur cycle (figure 1.1). Direct geochemical measurement of the pO_2 of ancient atmospheres has been attempted (Berner and Landis 1988), but the findings remain controversial (Cerling 1989). It is possible that pO_2 changes suggested by modeling (Berner and Canfield 1989) were of sufficient magnitude during the Phanerozoic to have significant ecological and even evolutionary effects (Graham et al. 1995). However, neither such changes in geochemical development nor the catastrophic events associated with the great mass extinctions produced outright gaps in the record of continuity of life on the planet.

Indeed, the evidence of the existence of living organisms on the surface of Earth over a period greater than 3 Gy places strong constraints on the outer bounds of variability of the global environment. In one of the early papers on Gaia, Lovelock and Margulis (1974) wrote,

> Stable temperature, pH, and element cycling requirements for life must have been met on this planet consistently for the entire 'recorded' history of the earth. Organisms found in extreme environments are highly specialized. If the earth had frozen out for even a few tens of thousands of years, or if hot acid had been widely distributed for even a single epoch, these occurrences would have been discerned from the fossil record.

The same point had been made some years earlier in a paper by Weyl (1966) entitled, "Environmental stability of the earth's surface—Chemical consideration." Weyl begins by observing,

> Life on the surface of our planet is possible because the environmental variables, the climate and the chemical composition of the atmosphere and hydrosphere, are suitable. The evolution of various forms of life during the last billion years requires in addition that these variables were stable during that time span. A variable is stable if, in response to a perturbation, the magnitude of the variable tends to return close to its original value. An unstable variable on the other hand will change radically if the equilibrium is disturbed. Without stability, life could exist temporarily if introduced, but it could neither persist nor evolve. Stability is relative. While minor climatic oscillations such as occurred during the Pleistocene may increase the rate of evolution, oscillations of much greater amplitude would lead to the extinction of life.

Weyl goes on to introduce the concept of a "life boundary." The set of n variables that characterize the environment can be thought of as defining an n-dimensional space, represented for a two-dimensional case in figure 3.2.

> The persistence of life requires that the environment be restricted to a bounded region in the environmental space. Let us call the boundary that encloses the maximum permissible excursions of the variables in the environmental space the "life boundary." Because of the mobility of organisms and the variety of the geography of the earth's surface the entire surface of the earth need not remain within the life boundary ... Because of evolutionary adaptability the life boundary is not

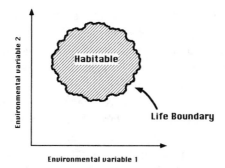

FIGURE 3.2 The "life boundary" concept introduced by Weyl (1966).

necessarily static... but can change through time, provided only that its rate of change is slow compared to the rate of biological evolution.

In summary, Weyl lists the mechanisms that may have been responsible for keeping the environmental variables within the life boundary. "These are: luck, size, chemical equilibria, and negative feedback." The following four sections will examine each in turn.

The Basis for Stability: (1) Luck

The considerations of chapter 1 provide one sense in which "luck" has played a part in the persistence of life on Earth. We saw there that forces arising from extraterrestrial collisions, tectonic cycling, or celestial mechanics have been responsible for extinctions, perturbations of the biogeochemical cycles, and glaciations. Had those forces been stronger, the survival of life on the planet might well have been in jeopardy. More fundamentally, we may inquire whether "luck" plays a part in situating the physico-chemical characteristics of Earth's environment within the life boundary in the first place.

The recent discussions of the "anthropic principle" (Barrow and Tipler 1986) suggest that observable properties of the universe such as age and size are functions of the time necessary for the evolution of planetary conditions appropriate to permit biological evolution. In 1913 Lawrence Henderson published his classic work, "The fitness of the environment," in which he discusses the question of what physico-chemical properties of the components of the environment are essential to make life possible (Henderson 1970). It is arguable that such cosmological considerations are mere truisms. The universe that we observe is, ipso facto, habitable. If other universes exist, they may or not be viviferous; but, since they are unobservable, no comment can be made on what makes for the ability to bear life. On the other hand, within the observable universe, there are many celestial bodies, including planets of this solar system, that are almost certainly sterile.

The contrast in this respect is particularly striking among the three rocky planets: Venus, Earth, and Mars. With Earth taken as 1, the relative mean distances from the sun are 0.72:1 for Venus and 1:1.52 for Mars. Venus and Earth are of similar mass (0.81:1) and equatorial radius (0.95:1), and Mars and Earth have nearly identical periods of rotation (1.03:1). The relative inputs of solar energy to the tops of their planetary atmospheres are 1.9:1:0.43 (Kivelson and Schubert 1986). Despite these relatively small differences, Earth has a flourishing biota, while Venus and Mars are sterile and probably always have been.

It appears that the solar/planetary conditions for habitability are relatively stringent. Venus is too close to the sun and therefore too hot; Mars is either too far from the sun or too small and therefore too cold; only Earth is "just right." It is not surprising that this has been nicknamed the Goldilocks theory of planetary habitability (Kasting et al. 1988). The possibility exists that "near-surface aqueous environments existed on Mars early in its history and could have provided conditions much more favorable for life than those that exist today" (Squyres and Kasting 1994). The possibility that life existed transiently on Mars but could not survive is an important consideration for any general theory of planetary environmental homeostasis.

Being at the right place in a solar system and being the right size may not be the only ways in which "luck" plays a part in determining planetary habitability. Fortuitous aspects of planetary chemistry may be important, particularly in determining whether or not an oxygen-rich atmosphere and hydrosphere develops, in turn permitting the evolution of complex life forms. Most of the oxidizing equivalents produced by the photosynthetic photolysis of water during geological history are not accounted for by the oxygen of the atmosphere. Most of the oxygen liberated during the Precambrian would have been recaptured by inorganic sinks, such as sulfide (S^{2-}) or ferrous iron (Fe^{2+}), (Broecker 1970a; Schidlowski et al. 1975). Today's global inventory of the oxidized products of the long history of photosynthesis is, indeed, dominated by sulfate and ferric iron (table 3.1). The transition to the oxygen-rich atmosphere of the Phanerozoic must have occurred as the reducing sinks became exhausted, although the stoichiometry of

The Life Boundary and Environmental Homeostasis 71

TABLE 3.1
Global Inventory of the Oxidized Products of Global Photosynthesis

	Oxygen Equivalents (10^{20} moles)
molecular oxygen (O_2)	0.4
sulfate (SO_4^{2-})	4.4
ferric iron (Fe^{3+})	1.6–3.2

NOTE: Sulfate and ferric iron are expressed as equivalent to the amount of O_2 consumed in their production.

liberation and recapture during this transition is not well understood (Van Valen 1971).

Garrels and Perry (1974) have pointed out that the nature of the ferrous sink on a planet will have important consequences for the persistence of life. If the sinks are predominantly ferrous carbonate (siderite, $FeCO_3$), the consumption of O_2 would be accompanied by the liberation, on a mole-for-mole basis, of four times as much CO_2. On the other hand, the oxidation of inorganic sinks such as pyrite (FeS_2) and ferrous silicate ($FeSiO_3$) would remove oxygen from the atmosphere without any release of CO_2. It is possible to write a balanced chemical equation for a photosynthetic process in which reducing equivalents are transferred from a mixture of FeS_2, $FeSiO_3$, and $FeCO_3$ to CO_2, thus producing organic carbon. The ferrous iron is oxidized to Fe_2O_3, and the pyrite sulfide to sulfate. The mixture of products also includes free O_2. The stoichiometry of the reaction is shown in figure 3.3, which demonstrates that the amount of oxygen produced is dependent upon the amount of $FeCO_3$ in the original mix.

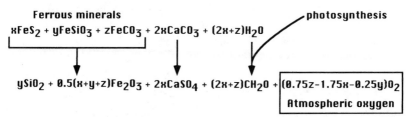

FIGURE 3.3 The stoichiometric relationship between the production of atmospheric O_2 and the primordial distribution of ferrous iron between sulfides, silicates, and carbonate. Note that the amount of O_2 in the atmosphere depends upon how much greater 0.75z is than the sum of 1.75x and 0.25y.
SOURCE: Garrels and Perry 1974.

The level of free oxygen in the present steady-state atmosphere would seem to be fortuitous, in that it apparently was controlled by the relative amount of siderite in the preoxygen rocks. If there had been more siderite, the Phanerozoic steady state might have operated at a somewhat higher level. (Garrels and Perry 1974)

The Basis for Stability: (2) Size

The second mechanism that Weyl (1966) lists as having contributed toward the maintenance of the environmental variables within the life boundary is "size." We saw in chapter 1 that different reservoirs of the biogeochemical cycles will have different turnover times, defined as reservoir size divided by flux rate (figure 1.9). In a steady-state cycle the flux rate will be the same for all reservoirs, and therefore the turnover time will be directly proportional to pool size. The half-life for response of reservoir size to a change in flux rate will be, in the simplest case, ln2 × turnover time (that is, $0.69 \times R/f$). The stability of the different compartments of a biogeochemical cycle to a change in flux rate is therefore strongly influenced by the size of the pools.

Size is of obvious importance in determining the effect of a sudden injection or withdrawal of material. The previous chapter explained that it is the quantitatively small pools of the minor atmospheric constituents that are most vulnerable to human impact. On the other hand, discussing the effects of meteorite impacts, Toon (1984) observed: "The quantity of oxygen and nitrogen in the atmosphere is so large that it is not likely to be altered significantly except by the degassing of a magma reservoir or impacting body of volume larger than 5×10^6 Km^3." For comparison, the volume of the bolide believed responsible for the extinctions at the Cretaceous/Tertiary boundary has been estimated at about 4×10^3 Km^3.

TABLE 3.2
Atmospheric Oxygen, 1910 to 1970

Year	O_2 (% by volume)
1910–12	20.952 ± 0.007
1919	20.948
1930s	20.938 ± 0.003
1935	20.946 (20.935–20.950)
1942	20.92
1967	20.946 ± 0.006
1970	20.946 ± 0.0018

SOURCE: Machta and Hughes 1970.

TABLE 3.3
Potential Sinks for Atmospheric Oxygen

	Half-Life (years)	Pool Size (Pmoles)
terrestrial biomass	5	<75
soil carbon	>200	<200
fossil fuels	<1400	<1000
sedimentary organic	100×10^6	1.3×10^6
pyrite	300×10^6	0.2×10^6
atmospheric O_2	2.6×10^6	38×10^3

NOTE: Half-life = 0.69 × (reservoir size) / (flux rate).

In the 1960s, when concerns about the global environment were first widely expressed, the possibility was raised of anthropogenic impact upon the world's supply of atmospheric oxygen. The knowledge that atmospheric oxygen is of photosynthetic origin continues to lead the public to the assumption that human impacts upon tropical forests or marine plant life may significantly affect the level of oxygen. High precision measurements of the oxygen content of the atmosphere have shown that there was no empirical foundation for this particular concern (table 3.2). Simple calculations based upon the size of the atmospheric pool of oxygen also show these fears to be misplaced (Broecker 1970b; Garrels et al. 1976). Today's atmosphere contains about 38,000 Pmoles O_2. This figure should be compared with the global carbon pools (table 3.3). Pyrite constitutes another major sink for oxygen.

It is clear from these figures that the only pools capable of reducing atmospheric oxygen on a timescale of human affairs are relatively small. The stability of atmospheric oxygen, on this timescale, is conferred by size and is such that, even in a "doomsday" scenario of complete cessation of photosynthesis, the prospect of anoxic asphyxiation would be low on the list of reasons for human concern.

The Basis for Stability: (3) Equilibrium

The third mechanism on Weyl's (1966) list of factors tending to stabilize the characteristics of the global environment is "equilibrium." A relatively small pool may be stabilized by being in equilibrium with a large pool that will act to buffer the smaller pool against perturbation. Thus, the relatively small pools of some atmospheric constituents may be stabilized by being in equilibrium with the much larger pools of the hydrosphere and lithosphere. Specific suggestions in this regard have been made. Urey (1952) suggested that the fundamental long-term control of the level of CO_2 in the atmosphere is exerted by the equilibria of quartz with the silicates and carbonates of magnesium and calcium in reactions such as,

$$CaCO_3 + SiO_2 \rightarrow CaSiO_3 + CO_2$$

and

$$MgCO_3 + SiO_2 \rightarrow MgSiO_3 + CO_2$$

This simplistic view has been superseded by detailed examination of the equilibria between clay mineral phases in the ocean. Sillén (1961) proposed that the composition of seawater was that of an aqueous solution in equilibrium with clay minerals and carbonates. The typical reaction may be represented as,

$$1.5\ Al_2Si_2O_5(OH)_4 + K^+ \rightarrow KAl_3Si_3O_{10}(OH) + H^+ + 1.5\ H_2O$$

The solid phases—kaolinite on the left and illite on the right—and water all have an activity of 1; therefore the equilibrium for this ion exchange reaction may be written $[H^+]/[K^+]$ = constant. Under the constraint of electroneutrality, at constant temperature and salinity, both $[K^+]$ and $[H^+]$ are determined. The partial pressure of CO_2 is a function of the sum of carbonate and bicarbonate ions and the pH. Therefore, if pH is fixed by this equilibrium, so will be the partial pressure of CO_2 in equilibrium with carbonate minerals. A more realistic version of the Urey carbonate/silicate equilibrium (Holland 1978) is therefore

$$CaCO_3 + 8SiO_2 + 7Al_2Si_2O_5(OH)_4 + 4H_2O \rightarrow$$
$$CaAl_{14}Si_{22}O_{60}(OH)_{12} \cdot 24H_2O + CO_2$$

Here, calcium carbonate, silica, kaolinite, and water react to form calcium-montmorillonite and carbon dioxide. Again, the activities of all reactants and products except CO_2 equal 1; at constant temperature and salinity, pCO_2 is also constant.

Experimental evidence for clay mineral equilibria has been difficult to obtain. Such reactions are slow to reach equilibrium. If they are important, their significance must lie in the control of very long-term processes. Sillén himself wrote,

This does not mean that I suggest that there would be true equilibrium in the real system. In fact practically everything that interests us in and around the sea is a symptom of non-equilibrium: the various forms of life, the currents, the shifting weather and so on. What one can hope is that an equilibrium model may give a useful first approximation to the real system and that the deviations of the real system may be treated as disturbances. (Sillén 1967)

The Basis for Stability: (4) Feedback

Weyl's fourth suggested mechanism for ensuring environmental stability is "negative feedback." In a mathematical sense, the effectiveness of negative feedback means that vectors in environmental space stay within the life boundary, that if the environmental variables are changed they will tend to shift back toward their original values, and the path to those values will not cross the life boundary. In a geochemical sense, negative feedback is equivalent to Le Chatelier's Principle, a general rule applicable to all chemical systems. The principle states that, if one of the variables that determine the state of a system is changed, the system will change in such a way that the magnitude of the change of the variable is diminished. The principle thus worded is quite vague and has been applied to a wide range of situations. In a rigorous form, based on the Second Law of Thermodynamics, where it receives expression as the van t'Hoff isotherm, its application is restricted to closed systems at equilibrium. Le Chatelier's Principle would not therefore suggest any basis for stability additional to that offered by equilibrium considerations (De Heer 1957, 1958).

If, however, the geochemical systems in question are in an open steady-state or approaching such a steady-state, in which rates of production of components of the system are balanced by rates of decay (though the components may be far from equilibrium with one another), then such steady-state systems will possess greater stability than will closed equilibrium systems. In one of the earliest analyses of this stability, Burton wrote,

> As they have been formally described, the steady-state systems already considered possess an unlimited stability. Suppose that a fluctuation occurs which increases the concentration of a reactant above its steady-state level. However large that fluctuation may be, the rate of transformation of that reactant will increase so as to restore it to the

normal level. This is the consequence of the law of mass action that is stated in general terms by Braun and Le Chatelier. (Burton 1939)

It is presumably some such understanding of Le Chatelier's Principle that Berger had in mind when, in criticizing the Gaia hypothesis, he wrote, "Disproof of this 'hypothesis' is hard to imagine as long as Le Chatelier's Principle is at work and as long as life is able to adapt to changing conditions on the surface of the planet" (Berger 1984). The second of these two conditions, the ability of the biota to expand the life boundary by effective biological adaptation, will be a major theme of chapter 5. The first condition, the existence of mass action control of the geochemical steady-state, is widely recognized. In this view, for instance, the chemical composition of the oceans is determined not by equilibrium with the lithosphere but by the balance between inputs from the rivers and atmosphere and removal to the sediments and to the atmosphere (Broecker 1971).

To return to the previous example, it may be the exchange of K^+ and H^+ between clay minerals and the aqueous phase that determines the concentration of these cations in natural waters. Nevertheless, it is important to recognize that the displacement of K^+ by H^+ must take place on the continents as part of the weathering process, while the adsorption of K^+ and concomitant release of H^+ ("reverse weathering") will take place in the ocean. It is likely that the end result of such a combination of processes will be determined by kinetic factors. The rate of weathering will influence the concentration of K^+ in the riverine input to the oceans; the kinetics of "reverse weathering" will, in this model, determine the rate of withdrawal of K^+ to the sedimentary sink.

Similarly, in the steady-state models pCO_2 would not be determined by the equilibrium constant for the reaction of CO_2 with calcium-rich clays to give calcium carbonate, kaolinite, and quartz. Instead, pCO_2 would be set by the balance between the rate of removal of CO_2 in the weathering process and the rate of recycling of sedimentary carbonates through volcanic outgassing.

The removal of atmospheric CO_2 through weathering provides an example of a negative feedback mechanism keeping global temperature stable within the life boundary. The Goldilocks theory, in which Earth's habitability is a fortuitous consequence of planetary size and distance from the sun, fails to account for the origin of life on Earth some 3.5 Gya. Astrophysical theory suggests that at that time the solar luminosity was only about 75% of its current value. Under such circumstances the mean temperature at the bottom of an atmosphere of present-day composition would be below 0°C. If, on the other hand, the composition of the Earth's early atmosphere had been such as to yield a surface temperature at which water is liquid, then, if that atmospheric composition had been retained, today's temperature might be outside the life boundary on the upper end (Kasting et al. 1988).

It has been suggested (Walker et al. 1981) that the resolution of this paradox lies in a feedback mechanism coupling surface temperature and pCO_2 in such a way that an increase in temperature (as solar luminosity increased) would be counteracted by a decrease in pCO_2 and, hence, a decreased greenhouse effect of that gas. The connection is probably mediated via rainfall and weathering. Increased temperature leads to increased evaporation from the oceans and thus to increased rainfall. Increased rainfall leads to increased weathering that would draw down CO_2, reduce the atmosphere's absorption of infrared radiation, and thus lower the surface temperature.

Because CO_2 is itself a determinant of habitability, it is possible to construct a diagram (figure 3.4) in which a two-dimensional space is defined by atmospheric pCO_2 and solar luminosity (the latter being expressed relative to the present-day value). The two isotherms are set at the freezing point of water and 100°C. They are drawn using the relationships among pCO_2 today and at life's origin (P and P_0), solar luminosity (L and L_0), and absolute temperature (T), (Walker et al. 1981).

$$P/P_0 = [49.04 + T/4.6 - 110(L/L_0)^{1/4}]^{2.747}$$

The vertical dotted line in figure 3.4 indicates the probable value of solar luminosity at 3.8 Gya when life originated. The downward sloping dashed line presents a possible trajectory for atmospheric pCO_2 since then. A more accurate representation of that trajectory would allow for the possibility of significant variability of atmospheric CO_2 in the past 600 My (Berner 1990), but the general argument would be unaffected. The lower bound is set somewhat arbitrarily to give a value of 130 ppmv CO_2 at present-day solar luminosity, which would decrease net primary production by >80%. Assuming plant respiration to be decreased by 50% for a 10°K decrease in temperature, then a similar limitation at temperatures approaching 0°C would occur at 70 ppmv CO_2; linear extrapolation gives the value of 12,000 ppmv at 100°C.

Lovelock and Whitfield (1982) have pointed out that, on such a geo-

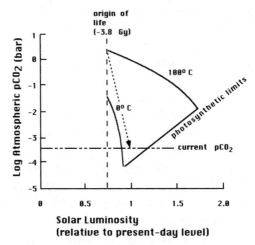

FIGURE 3.4 A "life boundary" in the two dimensions of solar luminosity and atmospheric CO_2. The zone of habitability is bounded by the 0°C and 100°C isotherms and by the lower limit of pCO_2 that permits net photosynthesis. The equation used to calculate the 0°C and 100°C boundaries is given by Walker, Hays and Kasting (1981). For "photosynthetic limits" see text.

logical timescale, the system is now approaching the lower limit of pCO_2 for photosynthesis. Limitation of global biomass, either marine or terrestrial, would cause CO_2 to increase, so a possible scenario is for pCO_2 to stabilize somewhat above the photosynthetic compensation point. Any further increase in solar luminosity would then be uncompensated, and the surface temperature would increase on a million-year timescale to a point at which the planet would cease to be habitable for all but thermophilic bacteria, and ultimately even for those forms. Somewhat different scenarios for the end of habitability in a low CO_2 atmosphere are discussed by Caldeira and Kasting (1992) and by Lapenis and Rampino (1993).

From a standard geochemical point of view, environmental parameters are determined by chemistry and physics. A combination of many factors—the size of the planet, the level of solar luminosity, the distance from the sun, the values of the constants characterizing mineral equilibria—has given rise to a planetary environment that is amenable to life. That such a system would be stable is to be expected from the general properties of chemical equilibrium expressed in Le Chatelier's Principle and, particularly, from the application of the Law of Mass Action to steady-state systems. From such a point of view, life is possible in such an environment but does not play any significant part in determining the parameters characterizing that environment. Thus, Holland (1984:539) writes that he believes the continuity of life to be "a consequence of the relative dullness of Earth history, of the rarity and relatively small magnitude of disruptive events such as asteroid impacts, of the variety of physical and chemical control mechanisms that have tended to maintain the status quo and to damp out rather than to amplify fluctuations." Similarly, in their discussion of the negative feedback loop between temperature and pCO_2, Kasting, Toon, and Pollack (1988) write, "There is good reason to believe that the earth would still have remained habitable even if it had never been inhabited. The carbonate-silicate cycle, acting alone, would have provided the necessary buffering mechanism."

A major difficulty with this point of view is that, in contrast to a system at equilibrium, the composition of a steady-state system is strongly influenced by catalysis of the processes that make up the system. It is undeniable that many of the processes of global metabolism are catalyzed by the biota. Many of the feedback mechanisms that have been postulated by geochemists actually include biological components at key junctures. Thus the biota are indeed implicated in and may even determine the characteristics of geochemical and geophysical feedback loops.

Biochemical Processes in the Geochemical Feedback Loops

For a component of a biogeochemical system in a steady-state, there must be some mechanism to ensure that the rate of formation of the component and the rate of its removal must be equal. In figure 1.9, for example, $f_{in} = f_{out}$. The mechanism may be simple, as in the case of a first-order system in which an increase or decrease in the component will correspondingly change the rate of removal until a new balance with the rate of formation is achieved. Or stabilization may involve feedback mechanisms of varying degrees of complexity.

Not only the stability but also the level of the steady-state will be influenced by the kinetic characteristics of the system. Kinetic effects on the steady-state level are an important property that distinguishes open systems from systems at equilibrium. In the latter, the equilibrium position of a chemical reaction is unaffected by the catalysis of the forward and backward components—that is, the rate constants for the opposing reactions do not enter explicitly into the calculation of the composition of the mixture at equilibrium.

Throughout the rest of this chapter, one key point will recur: *in the global biogeochemical cycles, some of the kinetic characteristics that determine both stability and steady-state values of some components of the global environment are attributes of particular biological species—and these attributes are determined at the molecular level.* That is to say, if one could write out a mathematical expression from which both the steady-state and the time-dependent behavior of these environmental components could be calculated, the expression would have to include terms that correspond to biochemical characteristics of the organisms involved in the global cycle. This dependence of global environmental characteristics upon biochemical processes, catalyzed by the global biota, provides much of the justification for the metaphor of global metabolism. Three examples—oxygen, carbon dioxide, and dimethyl sulfide—will be used to substantiate the concept that biochemical processes do play a consequential part in determining the stability and level of global environmental parameters.

Biochemical Controls: (1) Oxygen

The first case to be considered is that of atmospheric O_2. It was pointed out earlier that, on the timescale of human history, the level of atmospheric O_2 is stable (table 3.2) and that this stability is attributable to the size of the atmospheric reservoir of O_2 in comparison with the size of the sinks capable of reacting with O_2 on such a timescale (table 3.3). There are two ways in which this claim for stability must be modified, however. In the first place, if the seasonal cycles of atmospheric CO_2 that are a conspicuous feature of the Mauna Loa records (figure 2.1) are a consequence of the annual cycle of terrestrial photosynthetic production and respiratory decay, then these fluctuations in CO_2 should be accompanied by stoichiometrically equivalent fluctuations (but of opposite sign) in atmospheric O_2.

This expectation was pointed out a century and a half ago by the French scientist M. J. Dumas, who realized that the summation on a global scale of the metabolic activities of plants and animals should be reflected in the composition of Earth's atmosphere. In a lecture published in 1841, "On the chemical statics of organized beings," Dumas wrote,

> As animals breathe continually: as plants breathe under the solar influence only: as in winter the earth is stript, whilst in summer it is covered with verdure; it has been supposed that the air must transfer all these influences into its constitution. Carbonic acid should augment by night and diminish by day. Oxygen, in its turn, should follow an inverse progress. Carbonic acid should also follow the course of the seasons, and oxygen obey the same law.

Dumas goes on to anticipate the question of the stability of the atmosphere's composition. He recognizes size as a major factor in that stability:

> As to this, then we cannot be deceived; the oxygen of the air is consumed by animals, who convert it into water and carbonic acid; it is restored by plants, which decompose these two bodies. But nature has arranged everything so that the store of air should be such with relation to the consumption of animals, that the want of intervention of plants for the purification of the air should not be felt until centuries have elapsed.

Dumas believed this overall stability to hold both for oxygen and carbon dioxide, although he was aware of irregular fluctuations in measurements of carbon dioxide. The advent of modern analytical techniques has borne out Dumas' predictions. The seasonal fluctuations of pCO_2 measured at Mauna Loa, driven by the northern hemispheric terrestrial vegetation cycle, have become an icon of global metabolism.

Until recently the corresponding fluctuations of pO_2 lay below the limits of detectability. The removal of more than 4 Pmoles of CO_2 annually from the atmosphere by terrestrial photosynthesis must be accompanied by the addition to the atmosphere of an equivalent number of moles of O_2. The removal of CO_2 by northern deciduous plants occurs from a total pool of about 60 Pmoles and brings about an annual cycle that is relatively easy to detect. By comparison, the atmospheric pool of O_2 is 38,000 Pmoles; the annual fluctuations in response to seasonal changes in photosynthetic activity are correspondingly difficult to detect—an example of Weyl's point that size can provide a basis for stability. The technical difficulties have now been overcome (Keeling and Shertz 1992), and seasonal cycles of atmospheric oxygen with an amplitude of about 22 ppmv have been measured. To appreciate this achievement, recall that the level of atmospheric O_2 is about 21% by volume; the seasonal cycles are therefore only about $1/10,000^{th}$ of the background level. Because this signal can be used to separate the atmospheric effects of oceanic photosynthesis from those of terrestrial plant production, this technical breakthrough is going to be of great value in probing global metabolism. These quantitatively minor fluctuations do not affect the general claim for stability of the global atmospheric pool of O_2 on the timescale of human history (table 3.2). In fact, they provide our first quantitative measurement of the stability of that pool. There is, however, a second way in which the claim for stability of atmospheric O_2 must be qualified.

Table 3.3 shows that there are indeed very large sinks for O_2 on the much longer scale of geological time. A cessation of life on this planet would therefore be followed by a complete removal of free oxygen from the atmosphere by reaction with the massive reducing sinks of the lithosphere. The long-term stability of atmospheric O_2, which can be inferred from the continuous fossil record of aerobic fauna from about 600 Mya, cannot be explained on the basis of reservoir size.

It is generally accepted that the input of O_2 to the atmospheric reservoir is the small residue of annual photosynthetic production of O_2 that is not recaptured by the oxidation of the reduced products of photosynthesis. This input of O_2 is stoichiometrically equivalent to the amount of organic carbon (C_{org}) and other reducing materials, such as sulfides, buried annually on the sea floor. In turn, O_2 must be consumed in the oxidation of reducing materials in rocks undergoing weathering.

Holland (1978) estimated that the total sedimentary transport of rivers to the oceans is about 20 Pg/y. This figure can be used as an estimate of the total weathering rate and of total oceanic sedimentation. Knowing the C_{org} of marine sedimentary clay (given by Holland as 0.6%) and the C_{org}, S^{2-}, and Fe^{2+} of surface rock undergoing erosion (given by Holland as equivalent to 20g O_2/Kg), one can calculate the rates of input and consumption associated with the sedimentary cycle:

$$O_2 \text{ input} = 20 \times 0.006/12 = 0.01 \text{ Pmoles } O_2/y$$
$$O_2 \text{ used} = 20/32 \times 20 \times 10^{-3} = 0.0125 \text{ Pmoles } O_2/y$$

Garrels, Lerman, and Mackenzie (1976), employing the same approach but using somewhat different numbers, arrive at a figure of 0.0025 Pmoles O_2 per year. The four- or five-fold difference is irrelevant to the argument. The 37,500 Pmoles of O_2 in the atmosphere would have a turnover time of 3.75–15 My, a short period of time compared with the 600 My span of stability indicated by the fossil record. That stability implies a near equality between the input rate and consumption rate. The question then becomes one of proposing mechanisms that can maintain that equality.

It is reasonable to assume that both the rate of O_2 consumption (by weathering) and the rate of its production (stoichiometrically equivalent to the burial of reducing material generated by photosynthesis) will be affected by the pO_2 of the atmosphere. A number of authors have attempted to outline the shape of the functional dependencies of weathering on pO_2 and of burial on pO_2. The former is qualitatively straightforward. Oxidation by weathering involves the disintegration of rocks and the reaction of oxygen with the reducing materials exposed by this process. At low levels of pO_2 the rate of reaction with exposed reducing materials will be limiting. In contrast, at high levels of pO_2 the rate of exposure of the reducing materials will be the limiting factor, thus making the rate of oxidation of weathered materials independent of pO_2. One would therefore expect the kind of relationship typical of reactions that can be limited by the availability of reactive sites on a surface, i.e., the hyperbolic relationship of the Langmuir isotherm or the Michaelis-Menten equation (to be discussed in chapter 6). The quantitative aspects of this relationship are not well known. But it has been argued (Walker 1977) that the present-day pO_2 is sufficiently high that changes in its value would have relatively little effect on the rate of O_2 consumption by weathering. In other words, the present-day situation is located in the zero-order part of the relationship. A corollary of this argument is that the rate of O_2 consumption is determined by the erosion rate, which is itself changed by biological activity (as will be soon be shown).

The dependence of the burial rate upon pO_2 is less straightforward. The level of dissolved O_2 in seawater must play some part in determining how much organic material reaches the sediments unoxidized and is buried. Therefore, one may start by assuming that a decrease in pO_2 would increase the burial rate of undecomposed organic matter, correspondingly increasing the net photosynthetic production of O_2. Similarly, one might suppose that an increase in atmospheric pO_2 would raise the level of dissolved O_2 and thus ensure a more complete oxidation of organic detritus. This would lower the burial rate and correspondingly decrease the net photosynthetic production of O_2. The idea that the principal feedback mechanism stabilizing atmospheric O_2 relies upon the burial rate of organic carbon beneath the oceans (and hence on the concentration of O_2 in deep waters) has been widely accepted. But a recent summary of the empirical findings does not provide evidence for such a dependence (Holland 1990). In any case, it is certain that the relationship is more complex than indicated by this brief introduction.

In the first place, it is unlikely that the relationship between the oxidation rate of organic matter and dissolved O_2 is linear. Most of this oxidation occurs as a result of microbial activity. The terminal oxidases may be expected to have a high affinity for O_2, and so the oxidation rate will be independent of O_2 concentration down to very low levels. Organic matter will therefore escape oxidation only because it encounters severely hypoxic water or because the rate of sedimentation overwhelms the rate of oxidation. The latter effect is perhaps limited to highly productive shallow waters. It is important to recall that the estimated burial rate of C_{org} (10 Tmoles/y) is only about 0.3% of marine production. This means that the reoxidation of photosynthetic products is about 99.7% complete at the current level of pO_2 (0.21 atm). It is not obvious that even a doubling of pO_2 would reduce that rate to zero, at least by effects on oceanic respiration.

A second reason for the complex relationship between pO_2 and the oxidation rate turns on the fact that biological detritus contains more than just carbon. Biological detritus also carries essential nutrients, such as phosphorus and nitrogen. The burial of organic matter therefore removes these nutrients from ocean circulation and thus limits productivity in the surface waters. The loop,

$$\text{decreased } pO_2 \\ \downarrow \\ \text{increased burial of } C_{org} \\ \downarrow \\ \text{increased rate of net } O_2 \text{ production,}$$

is therefore limited by the simultaneous operation of another loop,

$$\text{decreased } pO_2 \\ \downarrow \\ \text{increased burial of nutrients} \\ \downarrow \\ \text{decreased rate of marine photosynthesis.}$$

We have already estimated that the current burial rate of C_{org} is about 10 Tmoles/y. The molar ratio of phosphorus to carbon (P:C) in marine phytoplankton is 1:106, but during the process of sedimentation this ratio is altered by a preferential loss of P so that the P:C ratio of the buried material is only half that of the original biomass. The 10 Tmoles of C_{org} deposited annually are therefore accompanied by $10/212 = 0.047$ Tmoles P/y. The riverine input to the oceans is estimated to be about 0.065 Tmoles P/y (table 1.1), so removal as organic detritus accounts for more than half of the total oceanic flux of P (Holland 1978). The oceanic pool of P is only about 2.6 Pmoles (table 1.1); therefore an efflux significantly in excess of the riverine input would rapidly deplete it. For instance, a four-fold increase of the sedimentary burial to 0.188 Tmoles/y would result in a net exit of 0.123 Tmoles P/y. Oceanic P would fall, with a half-life of $0.69 \times 2600/0.123 = 14{,}600$ y. Clearly, there are stringent constraints on the upper limit of C_{org} burial as organic detritus. It thus appears that the net photosynthetic rate of production of O_2 could not rise much above the current value of 10 Tmoles/y.

A third complicating factor for understanding the relationship between pO_2 and the oxidation rate arises from the fact that much of the organic matter that reaches the sea floor does so adsorbed on mineral surfaces—referred to by Lee (1994) as a sort of global "kitty litter"! As pointed out by Kell et al. (1994), the possibility that the erosion of terrestrial rocks might play a part both in oxygen consumption and in providing the substrate that carries photosynthetic product to the marine sediments suggests another explanation for the balancing of oxygen consumption and production rates.

All in all, considering the three complicating factors, it does not appear possible to draw with any confidence the exact graphical relationship between the level of atmospheric oxygen and its rates of consumption and production. Holland (1978) and Walker (1977) have each sketched such relationships; figure 3.5 recapitulates some of their suggestions. The curve representing O_2 consumption (weathering rate) has been drawn in accordance with Walker's argument that the weathering sink for O_2 is relatively independent of pO_2 at present-day values. In comparison with Holland's diagram, the O_2 consumption rate is shown in figure 3.5 with a somewhat higher "affinity" for O_2, so that the steady-state value (20% atm) falls in a region where the order of reaction is closer to zero. In that region the O_2 consumption rate is determined by the erosion rate. The curve representing net photosynthetic

86 The Life Boundary and Environmental Homeostasis

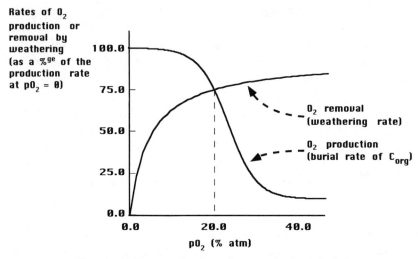

FIGURE 3.5 The rates of net production of oxygen by burial of photosynthetic products (organic carbon, C_{org}) and of consumption of oxygen by weathering as functions of the atmospheric level of oxygen. Notice the two curves intersect at today's atmospheric pO_2 of about 0.21 atm. This intersection represents a steady-state in which the production rate and the consumption rate are balanced. The stability of that steady-state depends upon the angle of intersection of the two curves. The greater that angle, the more stable is the steady-state.
SOURCE: Based in part on figure 3.2 of Walker 1977 and figure 6-12 of Holland 1978.

production of O_2 (burial rate of C_{org}) follows Holland's arguments with respect to the limitation of the rate by nutrient depletion at low pO_2. But for reasons already stated, it is assumed that some production might persist even at high O_2 levels.

For the purpose of this chapter, the extrapolation of the oxygen production and removal curves to low or high values of pO_2 (as shown in figure 3.5) is not important. What does matter is the shape of the curves near their intersection at a value of pO_2 close to 0.21 atm. That intersection represents a steady-state in which the production rate and the consumption rate are balanced. The stability of that steady-state depends upon the angle of intersection of the two curves. The greater the angle, the more stable is the steady-state. At a greater angle of intersection, a given displacement of the steady-state will have a larger effect on the difference between the rates. It is this difference that will act to restore the original steady-state. The relative slopes of the curves of figure 3.5 suggest that the response of the burial rate of photosynthetic products to pO_2 is the most important factor in the stability of atmospheric pO_2. The burial rate, in turn, is strongly influenced by such biochemical factors as the dependence of microbial respiration on concentrations of O_2 and the dependence of planktonic photosynthesis on concentrations of P_{inorg}.

Such biochemical factors may not only influence the stability of atmospheric O_2; they may also be involved in setting the value of the steady-state. How does it come about, in fact, that today's steady-state level of O_2 in the atmosphere is about 21%? The answer essentially requires an account of the value on the abscissa of figure 3.5 at which intersection occurs.

A variety of answers have been given to this question of locating the steady-state at that particular value. Earlier in this chapter, one of the examples given of "luck" playing a part in determining the habitability of the Earth was the suggestion, made by Garrels and Perry (1974), that the level of atmospheric O_2 was determined fortuitously by the composition of the Archaean lithosphere. Such a geochemical explanation provides no role for the biota in the determination of the Phanerozoic level of O_2.

There are, however, implicit biological assumptions built into other models. This is the case for the suggestion of Broecker (1970a) that atmospheric O_2 is maintained at a "kinetic maximum." Broecker accepts the general outline for the control of atmospheric O_2 just discussed. To reiterate, the principal controlling factor is the dependence of the rate of burial of organic detritus upon the level of oxygenation of the deep ocean. This concept requires that there must be oceanic zones in which O_2 is reduced to critically low levels. If such zones did not exist, burial of reducing equivalents would decrease, pO_2 would fall, and such zones would then come into being. On the other hand, if such hypoxic zones were larger, more reducing equivalents would escape oxidation and be buried, atmospheric pO_2 would increase in stoichiometric proportion to the increased burial, and, in turn, the hypoxic zones would shrink until a new steady-state is achieved. The level of O_2 at which stabilization occurs will therefore depend upon the details of oceanic ventilation, specifically the transfer of O_2 to deep waters. Other things being equal, the level of O_2 in the deep ocean would appear to depend upon the area and the depth of the global oceans. Because photosynthesis occurs at the surface, the rate of production of organic material is a function of area. Oceanic respiration is a function of both depth and area. In an ocean of area = A, of depth = D, of mean respiration rate (on a volume basis) = R, and of mean photosynthetic productivity (on an areal basis) = Prod, the burial rate will be given by the equation:

$$\text{burial rate} = A_m \times (\text{Prod} - R \times D)$$

For a given depth, the burial rate will increase with area. Therefore, the intersection with the consumption rate (figure 3.5) will shift to the right, to higher values of atmospheric pO_2. Conversely, at smaller areas (lower values of A_m) the hypoxic conditions that permit the burial of organic materials will be prevented by lower values of atmospheric pO_2 than would be necessary to prevent such hypoxia in a world ocean of larger volume. For hypothetical oceans of the same area, lower values of D will result in increased values of pO_2, a conclusion that is in accord with the greater burial of C_{org} in coastal areas (Walker 1977). Such an argument would suggest that the level of pO_2 at which balance is achieved between O_2 production and O_2 consumption is determined by the geophysics of ocean circulation. However, a calculation of the extent to which the intersection of figure 3.5 will be shifted for a given change in oceanic volume, even on the basis of the highly simplified assumptions built into the burial-rate equation, would necessitate knowing the dependence of the mean oceanic respiration rate on the concentration of dissolved O_2.

As mentioned earlier, microbial oxidation of organic detritus will be independent of O_2 down to low values. Put in biochemical terms, most bacteria have terminal oxidases (enzymes that catalyze the reduction of molecular oxygen) with Michaelis constants for O_2 in the order of a few µM. Richards (1957) reports that "in most aquatic organisms, respiratory consumption of oxygen is independent of the oxygen tension down to some lower limit of the latter." It is presumably this high affinity for O_2 that provides the biochemical explanation for the recent report (Holland 1990) that the burial efficiency of organic carbon is relatively independent of O_2 at the seawater-sediment interface. *The quantitative relationship of atmospheric pO_2 to ocean volume does therefore depend upon a particular characteristic of the biological processes involved, namely, the high affinity of the respiratory enzymes for O_2.* A higher value for the apparent Michaelis constant (affinity for oxygen) of the respiratory enzymes of marine heterotrophs would imply that the balance of weathering rate by burial rate would require a higher level of atmospheric O_2. That is, all else being equal, the point of intersection on figure 3.5 would be shifted to the right if the marine organisms decomposing deep-sea detritus were less well adapted to low oxygen levels. The explanation of the level of atmospheric O_2 in terms of the fortuitous composition of the Archaean rocks (such as given in Garrels and Perry 1974) may be called a geochemical account. The explanation in terms of oceanic volume (such as that given in Broecker 1970a) may be called a geophysical account. But Broecker's explanation is, in fact, incomplete without quantitative characterization of the biological processes that are implicit in the model. A full explanation of how it comes about that the steady-state level of pO_2 is now about 0.21 atm needs a biogeochemical account.

Such a biogeochemical account of this problem is given in Redfield's paper, the title of which, "The biological control of chemical factors in the environment," could well have served as title to this part of this chapter. As in all discussions of the stability and level of atmospheric O_2, the Redfield paper recognizes that the source of O_2 corresponds to the burial of the reduced products of photosynthesis. Like Holland, Redfield stresses that this deposition on the seafloor will be accompanied by the removal of limiting nutrients, such as N and P, with emphasis being laid upon phosphorus because the large atmospheric pool of N_2 makes possible replenishment of marine nutrient nitrogen by biological fixation. However, whereas in Holland's model the rate of O_2 production at 0.21 atm O_2 is limited by the efficiency of reoxidation of C_{org} (nutrient limitation becoming important only at low pO_2), Redfield argues that "the partial pressure of oxygen in the atmosphere [is] determined through the requirements of the biochemical cycle, by the solubility of phosphate in the ocean." Redfield's biogeochemical explanation is strikingly precise. Is it correct?

More recent discussions of the removal of phosphate from the ocean (Moody et al. 1981) suggest a more complex process than the simple precipitation of apatites (calcium phosphates) suggested by Redfield. In any case, the limitation of marine photosynthesis by the mean concentration of inorganic phosphorus ($[P_{inorg}]$) would appear to indicate only an upper bound for atmospheric O_2. However, Redfield suggests that $[P_{inorg}]$ would be held close to its solubility limits (and, consequently, atmospheric O_2 would be held close to 0.2 atm). This is because at low pO_2, when the oxidation of C_{org} by O_2 would be limited, oxidation by SO_4^{2-} (see chapter 1) would become the significant mode by which C_{org} burial would be prevented and nutrients such as P_{inorg} regenerated. In this model the rate of net photosynthetic production of O_2 would be relatively independent of atmospheric pO_2. The angle of intersection of the curves of figure 3.5 would be small, and both stability and the steady-state level of atmospheric O_2 would be determined by the dynamics of the marine biogeochemical cycle of phosphorus.

Holland (1990) states, "The marine geochemistry of phosphorus now seems to be the best candidate for the long-sought pO_2 control mechanism." And the missing link in a feedback loop controlling atmospheric O_2 may recently have been found (Van Cappellen and Ingall 1996). Phosphate is retained in oxic sediments and returned to the water column from anoxic sediments. More oxygen leads to less nutrient phosphate and therefore less photosynthetically derived organic carbon reaching the ocean floor. Conversely, lower pO_2 at the ocean floor leads to a liberation of phosphate, and so to greater surface oceanic production, and to an increased burial of organic carbon.

Note that in this, as in all variants of Redfield's hypothesis, the marine *geochemistry* of phosphorus is intimately bound up with the *biochemistry* of that element. We have seen that attempts to model Broecker's geophysical explanation would have to take into account the dependence of the respiration rate of deep sea bacteria upon O_2 concentrations. Similarly, any attempt to model Redfield's biogeochemical explanation quantitatively would have to take into account the dependence of the photosynthetic rate of marine phytoplankton upon P_{inorg} concentrations.

Marine phytoplankton are capable of photosynthesis at low concentrations of P_{inorg} (Chisholm and Stross 1976), the concentration in surface waters falling in the submicromolar range. This capability is attributable to active phosphate transport mechanisms, translocases in the cell membrane (Ducet et al. 1977; Ullrich-Eberius et al. 1981). There is therefore, in this model, a causal link between the affinity for P_{inorg} and the turnover of the phosphate translocases and the level of atmospheric O_2. If, for instance, the apparent Michaelis constant for P_{inorg} were higher, the global rate of O_2 production would be lower, as would be the steady-state atmospheric pO_2.

Two points that will become important as the arguments in this book are developed need emphasis here: (1) In both the geophysical and the biogeochemical explanations of the level of atmospheric O_2 and its stability, the control of this key environmental parameter by biochemical processes provides justification for the Gaian metaphors of global metabolism and geophysiology. (2) On the other hand, overinterpretation of such metaphors is strongly constrained by the recognition that the possession by phytoplankton of a molecular toolkit of enzymes and translocases (and the functional characteristics of these proteins) are phenotypic expressions of the genetic capabilities of these organisms, which were surely selected at the organism level by Darwinian evolution.

In other words, Gaian metaphors that connote a global physiology are helpful in portraying the crucial role of biota in the planetary cycles. But the heuristic value of these metaphors does not necessarily imply the much stronger contention of the originators of the Gaia hypothesis that life processes of global importance were selected *for* their contribution to global metabolism.

Biochemical Controls: (2) Carbon Dioxide

The second example that I will use to instantiate the way in which global environmental parameters are set by the kinetic characteristics of biological processes is the case of atmospheric CO_2. Earlier in this chapter it was suggested that a geochemical feedback loop existed in which the effect of an increasing input of solar radiation on global temperature was counteracted by a decrease in atmospheric CO_2. The critical process in this abiotic mechanism is weathering.

The reaction of CO_2 (or, more generally, of H^+ released by the deprotonation of H_2CO_3) with rocks is generally assumed to be limited by mechanical and hydrological factors (Silverman 1979; Lasaga 1984). Any increase in temperature and the consequent increased rainfall would thus feed back negatively on the CO_2 in the atmosphere by accelerating the reaction of rocks with CO_2. A more reactive surface would lead to a decrease in CO_2 and a correspondingly decreased greenhouse effect. There are, however, two ways in which the operation of this loop is interactive with biological processes.

First, the process of weathering is itself markedly accelerated by biological activity. Schwartzman and Volk (1989) have attempted to quantitate this effect in a paper entitled, "Biotic enhancement and the habitability of Earth." They list a number of ways in which mineral dissolution has been reported to be affected by biological activity. For example, the respiration of soil biota (including roots) generates a soil pCO_2 that is considerably higher than atmospheric pCO_2. Soil is itself formed and stabilized by biotic activity; in the absence of soil, rain water will not be held in contact with mineral surfaces. In the presence of soil, rain water is retained, and the time available for dissolution of mineral grains is considerably increased. Soil biota also enhance weathering by producing a variety of chemicals, such as humic acids, which assist in the attack on soil minerals and shift the equilibria in favor of soluble material. From data in the literature on the rate of weathering of bare rock compared to the rate of weathering in soils, Schwartzman and

Volk conclude that although "the enhancement factor is uncertain, the key role of soil stabilization combined with other effects suggests factors of the order of at least 100 to perhaps > 1000."

Figure 3.6 uses the equations of Schwartzman and Volk to calculate the behavior of the atmospheric feedback loop that may have been responsible for buffering global temperature against the long-term increase in solar luminosity (cf. figure 3.4) The two solid lines are the solutions to the simultaneous equations relating surface temperature to solar luminosity and to pCO_2, and relating the steady-state of atmospheric CO_2 in the sedimentation-erosion cycle to surface temperature (as in the isotherms of figure 3.4). The lower solid line uses rate constants for erosion characteristic of the present-day; the upper solid line is drawn assuming that, under abiotic conditions, weathering would be diminished by a factor of 1,000. At some period in global history a transition must have occurred from the abiotic "thermostat" controlling the value of the steady-state to the biologically controlled thermostat. The biologically controlled thermostat entails a more efficient removal of CO_2, which results in a lower steady-state level of atmospheric CO_2 and lower planetary temperature. The dotted line of figure 3.6 is drawn similarly to that of figure 3.4, to accord with recent discussions of global thermal history. It does not attempt to exhibit the fluctuations in pCO_2 that may have been characteristic of the most recent 600 My (Berner 1990).

The second way in which there is interaction between the biota and the CO_2-mediated feedback loop regulating global temperature arises from the dependence of the photosynthetic rate upon CO_2. The response of plant photosynthesis to changes in CO_2 levels is complex (Lemon 1983). At the enzymic level, the rate of incorporation of CO_2 is affected not only by the intracellular CO_2 concentration but also by the rate of supply of ribulose-1,5-bisphosphate (RuBP) and by competition with O_2 (photorespiration). For the so-called C-3 plants, the first enzymic reaction of CO_2 is catalyzed by ribulose bisphosphate carboxylase (Rubisco). Of the enzymes that catalyze the transition from the inorganic world to the world of cells and organisms, Rubisco is perhaps the most important. Rubisco may well be the most abundant protein on Earth. We saw in chapter 1 that it is not however perfectly adapted to its role. It catalyzes not only the reaction of CO_2 with RuBP but also a reaction with O_2. This reaction appears to achieve no biological purpose. Competition with O_2 increases the K_M (Michaelis constant) for CO_2 from 15 µM under anaerobic conditions to 26 µM in air (corresponding to a pCO_2 of 910 ppmv) (Tolbert and Zelitch 1983). At present-day levels of atmospheric

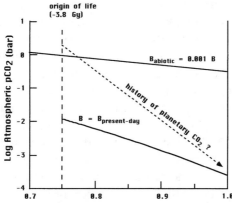

FIGURE 3.6 The influence of the biota on the relationship between solar luminosity and atmospheric CO_2. The equations in Schwartzman and Volk 1989 are solved for the abiotic case (upper trend) and for a case in which weathering is influenced by the biota as on the contemporary Earth (lower trend). For a defense of the view that the value of B for the two cases would differ by a factor as large as 1,000 see Schwartzman 1993.

CO_2, water contains about 10 µM CO_2, thus marked undersaturation obtains. At the cellular level, it appears that RuBP levels can become limiting at CO_2 levels as low as 6–7 µM, so that the response curve of photosynthesis flattens above this value (Caemmerer and Farquhar 1981).

In the later-evolved C-4 plants (many grasses and weeds are in this group), CO_2 is fixed first into malate or aspartate, which is transferred to the bundle sheath cells. Here, the process of decarboxylation maintains a high level of CO_2 at the chloroplasts—as much as 10× the level in C-3 plants (6 µM). CO_2 is thus able to compete more effectively with O_2, and wasteful photorespiration is suppressed.

Moving to the physiological level, the effects of differing values of CO_2 concentrations become more complex. Pearcy and Björkman (1983) list five areas of ignorance "in which more knowledge is required if we are to obtain a sound basis for predicting the response of plants to long-term effects of more CO_2." There are questions concerning the differing effects of increased CO_2 on different species, on water use and stomatal conductance generally, on leaf growth, on the way in which increased photosynthate is allocated among plant organs, and on the effect of other nutrients. If so many unresolved questions exist concerning the effects of CO_2 at the level of the individual plant, it is not surprising that when we move to the ecological level we find that the effects of CO_2 on natural plant communities are not well understood. It is known that under controlled conditions, as in greenhouses, crop growth is increased by elevated CO_2 (Bassham 1977). Because such increase occurs under otherwise optimal conditions of light, temperature, and nutrient supply, it is difficult to extrapolate from these results to natural ecosystems. Clearly, plants growing on soils that are poor in mineral nutrients may be unable to respond to increased atmospheric CO_2 (Oechel and Strain 1985). Even on fertile soil, the effects of increased CO_2 on plant growth may be modified if the transfer of photosynthate to the rhizosphere stimulates the growth of soil microorganisms that can compete effectively with the plants for mineral nutrient (Diaz et al. 1993).

The issue is much discussed recently (e.g., Mellilo et al. 1993; Dixon et al. 1994) because of the possibility that increased growth of the global vegetation in response to the anthropogenically driven increase in CO_2 (figure 2.1) could act as a significant sink for some of that CO_2. Most detailed models of the response of the global carbon cycle to the post-1850 input of CO_2 from fossil fuels and land clearance have used an increase in photosynthetic productivity to balance the budget between CO_2 inputs, the observed atmospheric accumulation, and the calculated uptake by the oceans (Bacastow and Keeling 1973; Broecker et al. 1979). An increase in carbon uptake by the terrestrial vegetation of the Northern Hemisphere is a conspicuous feature of the model of Tans, Fung, and Takahashi (1990). However, plant ecologists have been reluctant to accept these accounts (Woodwell et al. 1978). Chapter 2 noted the current uncertainties about the ocean uptake of CO_2. These uncertainties are large enough to call into question any estimates of the global increase in carbon uptake by the biota that are arrived at merely as a device to balance the global carbon budget. We saw that the deconvolution of the ^{13}C tree ring record (figure 2.4) could, in principle, be used to infer recent net changes in global biomass, but such deconvolution cannot be used to distinguish changes due to changed agricultural practices, forest clearance, or global eutrophication from those caused by "CO_2 fertilization." A major challenge for the future is to reconcile long-term ecological studies of the response of the biota to increased CO_2 with global oceanographic and atmospheric data.

The relationship of photosynthetic rate to pCO_2 is of interest, not only in connection with the contemporary increase in atmospheric CO_2 but also in connection with the geological history of this greenhouse gas. In figures 3.4 and 3.6, the dotted lines indicate a possible trend in atmospheric pCO_2 over the last 4 Gy. If this trend is correctly shown, then CO_2 is becoming increasingly limiting for plant growth. It has been suggested that the evolution of C-4 plants, with their higher affinity for CO_2, occurred in response to this decline in atmospheric CO_2 (Ehleringer et al. 1991). Others have attempted to relate the relative abundance of C-3 and C-4 plants to fluctuations of pCO_2 on the My timescale (Morgan et al. 1994) and on the Ky timescale (Cole and Monger 1994). But, for both C-3 and C-4 plants, there is a lower limit of pCO_2 below which no net photosynthetic fixation of CO_2 can occur. This limit is referred to by plant physiologists as the "compensation point" at which photosynthesis is balanced by photorespiration. For ecological purposes, the critical level is that at which net productivity falls to zero. This is the point at which gross primary production is balanced by dark respiration. Lovelock and Whitfield (1982) suggest a figure of 150 ppmv CO_2 as the lower limit tolerable for photosynthesis. A more conservative value for this threshold of 130 ppmv was used to draw the lower limits of the life boundary of figure 3.4. If C-4 plants were to dominate global ecology, this figure would doubtless be lower—though probably not as low as the figure of 10 ppmv CO_2 suggested by Caldeira and Kasting (1992), who seem to have used the value of the physiological compensation point for C-4 photosynthesis which takes into account carbon losses due to photorespiration but not those brought about by "dark" respiration.

It is of interest to note that this lower limit of CO_2 for net photosynthesis is actually approached by some of the lower values observed in ice core bubbles that correspond to the onset of periods of glaciation (Barnola et al. 1987). Clearly, even if the curve of pCO_2 drawn in figures 3.4 and 3.6 represents correctly the trend of atmospheric CO_2 levels since the origin of life 3.8 Gya, there may have been important deviations from that trend (note that the plot is semi-logarithmic). Such deviations during the past 150 Ky are recorded in the ice core bubbles (figure 1.2). For the past 14 Ky, at higher resolution, deviations are recorded in the $^{13}C/^{12}C$ ratio of peat deposits (White et al. 1994). If downward changes in CO_2 are triggers for global cooling, there must be some instability in the CO_2 thermostat. The changes are much too rapid to arise from the sedimentation-erosion cycle, which provides the basis for the control of global temperature shown in figures 3.4 and 3.6. The turnover time of the sedimentary rock pools is >100 My. On the other hand, the turnover times for the terrestrial biota and soil carbon are shorter by many orders of magnitude.

Remember that the set point of the long-term sedimentary feedback mechanism controlling the steady-state value of CO_2 was fixed by biological activity at a value considerably lower than would be expected for an abiotic Earth (figure 3.6). Is it biological activity which also generates the instability? At first glance this appears unlikely. A decrease in CO_2 would, as just discussed, reduce global productivity; thus the fall in CO_2 would be counteracted. Although the pools of terrestrial biomass and soil C (table 3.3) are sufficiently large that significant shrinkage or expansion could significantly affect atmospheric CO_2, the changes associated with glaciation or deglaciation would, at first glance, appear to be in the wrong direction. A decrease in production (productivity × biomass) moving into a glacial phase would leave more CO_2 in the atmosphere—the opposite of the ice core record. A decreased productivity might, however, be compensated for by an increase in land area due to sea level decline as ice volumes grow. Prentice and Fung (1990) conclude that "Although biospheric changes do not dominate the 82 ppm (165 Gt carbon) difference between the Holocene and 18 Ky BP found in ice cores, their contribution may not be negligible in calculating the components of large compensating factors in the biosphere-atmosphere carbon flux." Adams et al. (1990) similarly suggest that "Instead of contributing to the lower CO_2 level during the Last Glacial Maximum, the terrestrial vegetation and soil carbon reservoirs should be seen as a factor tending to 'damp' the oceanically driven glacial-interglacial CO_2 fluctuation."

These authors point out that this negative feedback is indirect, depending upon the response of plants to climatic factors rather than upon the relationship of photosynthetic rate to pCO_2. Such an indirect effect may also operate as a positive feedback, exacerbating the greenhouse effect of natural or anthropogenically forced changes in atmospheric CO_2 (Foley et al. 1994). Jenkinson, Adams, and Wild (1991) have estimated the effect of an increase in temperature on the decay of plant debris and other organic carbon of soil. The resultant release of CO_2 (estimated as 19% of fossil fuel sources) would further enhance the radiative forcing of global temperature.

The remaining important pool of inorganic C exchanged with the atmosphere is dissolved in the ocean. This dissolved inorganic carbon (DIC) is the sum of $[CO_3^{2-}] + [HCO_3^-] + [CO_2]$. The turnover time of oceanic DIC is about 200 Ky. This is too slow for DIC to be involved in the glacial-interglacial transitions. But the turnover between deep seawater and surface water is much shorter, many hundreds but perhaps less than a thousand years (table 1.1). It is equilibrium with the surface waters that controls atmospheric CO_2 on that timescale because atmospheric CO_2 is in rapid equilibrium with surface waters of the ocean. If the DIC of the surface waters were to fall, so would atmospheric CO_2.

The DIC of surface waters is lower than the average oceanic value (figure 1.11). We have seen that CO_2 is fixed into biomass and then falls from the photic zone as organic detritus. When this organic carbon is reoxidized, DIC is added to the deep sea. Marine photosynthesis in effect pumps DIC from the surface to the deep sea. An increase in marine productivity would therefore cause a decrease in atmospheric CO_2 and a fall in global temperature. Conversely, a decrease in marine productivity would permit a rise in DIC, as it is returned to the surface by ocean circulation; atmospheric CO_2 would rise correspondingly.

As was pointed out in chapter 1, marine productivity is responsible for the difference in ^{13}C content between the fossils of surface phytoplankton and those of benthic organisms. Suppression of photosynthesis will eliminate this difference. The sedimentary record does show such an effect at the Cretaceous/Tertiary boundary. It has been inferred from such evidence that a decrease in marine productivity was one of the consequences of a bolide impact (Zachos et al. 1989) and that a subsequent increase in atmospheric CO_2 and attendant rise in temperature must be included in the account of the mass extinction at that time (Hsü and McKenzie 1985).

As noted in chapter 2, anthropogenic emissions of CO_2 may be partially compensated for by oceanic eutrophication. Similarly, the decrease in pCO_2 at an interglacial-glacial transition might be attributed to an increase in planktonic photosynthesis brought about by an increase in a limiting nutrient, such as phosphate (Broecker 1982) or iron (Martin et al. 1990) or zinc (Morel et al. 1994). In contradistinction to this suggestion, Mortlock et al. (1991) have presented evidence of reduced accumulation of diatomaceous silica in Antarctic sediments during the last glacial period, suggesting that high oceanic productivity cannot be adduced as the cause of the decrease of atmospheric CO_2 in glacial periods. Any changes in the efficiency of the "biological pump" might also arise from changes in ocean circulation of nutrients. It thus is possible that changes in circulation would produce regional variations in productivity, different from those that exist today.

As noted in chapter 1, large temperature shifts (5–7°C) may have taken place during glacial-interglacial transitions in a matter of just 10–50 y (Dansgaard et al. 1989). Because it is not possible for ocean chemistry to adjust so suddenly, such changes are taken as evidence for profound, rapid changes in ocean circulation (Broecker 1987; Broecker and Denton 1989; Stocker and Wright 1991). The interplay of the physics of ocean circulation, the chemistry of the ocean carbonate system and the biology of marine phytoplankton is intricate. For example, a change in the ratio of calcite ($CaCO_3$) to C_{org} in the "detrital rain" would change the pH of the deep ocean in such a way as to strengthen the oceanic sink for CO_2 (Archer and Maier-Reimer 1994). Much further research will be needed to elucidate the way in which factors such as these have interacted to destabilize the feedback loops responsible for regulation of global temperature (Boyle 1990; Kudrass et al. 1991).

Biochemical Controls: (3) Sulfate and Clouds

The third example of an interaction between biochemical processes and the regulation of a global environmental parameter is the proposed relationship between global climate and the production of dimethyl sulfide, $(CH_3)_2 \cdot S$, or DMS, by marine algae. Charlson, Lovelock, Andreae, and Warren made this now much cited proposal in 1987. The posited interaction between biogenic sulfur compounds and climate is of particular significance because it has been used by Lovelock (1988) to substantiate the strongest version of the metaphor of global metabolism—namely, his own suggestion that the biogeochemical systems of planet Earth constitute the internal workings of a mega-organism, "Gaia."

It was Lovelock, Maggs, and Rasmussen (1972) who detected DMS in seawater and suggested that this biogenic compound could constitute a major component of the flux of sulfur from the oceans to the atmosphere. Only much later was it possible to measure the magnitude of this flux by direct measurements of DMS in the atmosphere. This is because DMS is present in the atmosphere at only very low concentrations, <100 pptv. By 1986 technical advances had made it possible to measure the vertical distribution of DMS in the atmosphere and to infer from that distribution a global input into the atmosphere of about 0.7 Tmoles/y (Ferek et al. 1986).

Dimethyl sulfide is produced by enzymic cleavage of dimethylsulfoniopropionate, $(CH_3)_2 \cdot S^+ \cdot CH_2 \cdot CH_2 \cdot COO^-$ (DMSP). The cleavage products are DMS and acrylate (Challenger 1959). DMSP plays an important role in the biochemical adaptation of marine algae to a saline environment (Dickson et al. 1980; Vairavamurthy et al. 1985). Chapter 5 will explore the idea that one of the most important modes of biochemical adaptation is the conservation of macromolecular function by the generation of an appropriate microenvironment. Adaptation to a hypertonic environment is commonly achieved by the metabolic production of an organic osmolyte, which must also be compatible with the maintenance of functions such as enzymic activity. In the course of evolution DMSP appears to have been selected as the compatible organic osmolyte of marine algae. The production of DMS from DMSP may be predominantly associated with the capture and ingestion of marine algae by zooplankton (Dacey and Wakeham 1986). DMS is removed from the atmosphere by oxidation, predominantly by OH. The two principal products of oxidation are SO_2 (>80%), and methyl sulfonate (MSA), $CH_3 \cdot SO_3^-$. The former is rapidly oxidized to SO_4^{2-}, while MSA is broken down to sulfate and CO_2 (Baker et al. 1991).

The possibility of climatic significance for this process arises because the sulfate and methyl sulfonate produced by the atmospheric oxidation of DMS form aerosol particles that can act as cloud condensation nuclei (CCN). The particles formed in this manner appear to be of more significance in cloud formation than those formed of sea salt. Cloud formation, in turn, can affect the global radiation budget. Charlson et al. (1987) calculate that a 30% change in the density of cloud droplets over the oceans would, if all other variables were held constant, change the global mean surface temperature by 1.3°C. This is a change of the same order of magnitude as the anticipated increase in global temperature between 1980 and 2030 due to the emission of greenhouse gases.

The connection between biogenic DMS emissions and global cloud albedo is not, however, accepted by all meteorologists. Schwartz (1988) has pointed out that the anthropogenic emissions of SO_2 are, on a global basis, more than twice the marine DMS emissions (cf. table 2.2). The anthropogenic emissions occur largely from the Northern Hemisphere. If the biogenic emissions can control cloud formation, then the effects of the larger anthropogenic emissions should be observable as a difference in albedo due to clouds and/or a differential history of temperature change between the two hemispheres. Because Schwartz was unable to detect any such differences, he concluded that the available records were inconsistent with the hypothesis of control of global albedo by marine algal production of DMS.

In contradistinction, Savoie and Prospero (1989) find that the biogenic production of DMS accounts for about 80% of the non–sea salt

SO_4^{2-} over the open ocean. Two other groups (Ayers et al. 1991; Prospero et al. 1991) have reported a strong seasonal correlation between atmospheric DMS and non–sea salt SO_4^{2-}, and between aerosol MSA and CCN concentration (Ayers and Gras 1991). A number of objections have been made to Schwartz's test (Henderson-Sellers and McGuffie 1989; Charlson et al. 1989; Gavin et al. 1989; Ghan et al. 1989; cf. the rebuttal by Schwartz 1989). One important consequence of this debate has been the recognition of a possible effect of SO_2 emissions on climate. Anthropogenic emissions of SO_2 may, by producing CCNs, be affecting global temperature in the opposite direction from the effects of anthropogenic increases in greenhouse gases (Hansen and Lacis 1990). It is even possible that, "if we were successful in halting or reversing the increase in SO_2 emissions, we could, as a by-product, accelerate the rate of greenhouse-gas-induced warming, so reducing one problem at the expense of increasing the rate of onset of another" Wigley (1989). (For recent general discussions of this problem see Jones, Roberts, and Slingo 1994, Charlson and Wigley 1994, and Mitchell et al. 1995).

The significance of DMS in the determination of climate is supported by measurements of MSA, which is an unequivocal indicator of DMS production, in Antarctic ice cores. The cores reveal an increase in the concentration of MSA (and non–sea salt SO_4^{2-}) during glacial times (figure 3.7). The low temperatures of the last ice age are associated with high MSA and therefore, by inference, with high DMS emissions, high density of CCNs, and high cloud albedo. This effect would reinforce that of the decrease in CO_2 associated with glacial periods. Both effects presumably reflect high marine productivity simultaneously enhancing the production of DMS and decreasing surface ocean concentrations of DIC. In both instances, the effect of the biota appears to have been destabilizing—perhaps a serious difficulty for the Gaian metaphor (Kirchner 1990). A drastic fall in marine productivity such as might be imagined to follow a major bolide impact (and as indeed is evidenced by ^{13}C measurements at the Cretaceous/Tertiary boundary) would bring about a decrease in DMS emissions (Rampino and Volk 1988) and in global cloud albedo. Once again, the effect of changed surface ocean productivity on atmospheric CO_2 would act synergistically with the DMS effect to bring about, in this case, an increase in global temperature.

Although the DMS-cloud-climate relationship provides a further instance of a biological process playing a significant role in the determination of a key global environmental characteristic, it does not appear to demonstrate negative feedback. Indeed, one of the problems

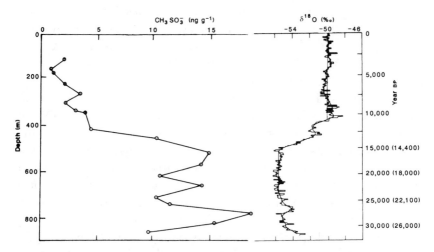

FIGURE 3.7 Evidence for increased atmospheric DMS in the last glacial period, drawn from Antarctic ice cores. (For an extension of this record see Legrand et al. 1991.)
SOURCE: Figure 2 of Saigne and Legrand 1987. Copyright © Macmillan Magazines Ltd. 1987. Reprinted with permission from Nature, Macmillan Magazines Ltd.

with using the production of DMS by marine algae as support for the Gaian metaphor is that it is unclear how negative feedback would work. Charlson et al. (1987) suggest that the supply of DMS and its oxidation products to the adjacent continents would increase terrestrial productivity and weathering and thus increase the riverine input of nutrients to the marine algae. Another suggestion is that the role of DMS in cloud formation would bring about greater open-ocean rainfall, which would wash out nutrients such as NO_x or terrestrial dust from the atmosphere. These ideas seem to suggest positive feedback loops and, in any case, do not connect in an obvious way to climate control. Bates, Charlson, and Gammon (1987) suggest a seasonal fluctuation in which low DMS production is associated with low incident solar radiation. The direction of the change is suggestive of a negative feedback mechanism, but the significance of such an effect either for phytoplankton ecology or for climate control is obscure. A further complication arises from the finding that there are significant sinks for DMS in seawater other than efflux to the atmosphere (Kiene and Bates 1990).

The three examples of "biochemical controls" given in this chapter have all concerned atmospheric constituents. The three, however, vary greatly in their abundance, from O_2 (which constitutes 21% of the atmosphere) to CO_2 (present at 340 ppmv) to DMS (present at <0.1 ppmv). All three gases are products of biological activity. Oxygen and CO_2 are undoubtedly of great significance for the habitability of Earth, and it appears that DMS may be also. These examples illustrate the subtle interplay among the composition of the global atmosphere, the global energy budget, and the biota that have evolved upon this planet. One of the most imaginative ways to think of this interplay is to liken it to the interactions that take place among the organ systems of complex metazoan organisms. Such an approach can lead to a view of the planet as a "superorganism." The study of such a planetary organism may appropriately be termed "geophysiology," a neologism derived by James Lovelock (1986) from an eighteenth-century suggestion of Hutton (1788). The next chapter, relying heavily on the ideas of Lovelock and Margulis, considers the utility of the geophysiological approach to the problems of environmental stability and global habitability.

4
Global Metabolism, Geophysiology, and Gaian Metaphors

In the previous two chapters I have occasionally made use of the construction, "global metabolism." What does it mean to speak of global metabolism?

Metabolism is understood by biologists as the orderly sequence of chemical transformations by which, for example, the carbon atoms of foodstuffs (such as starch) are used and decomposed into excretory products (such as carbon dioxide), are incorporated into storage products (such as fat deposits), or are made part of the cellular fabric of the metabolizing organism. The most obvious sign of an animal's metabolism is the rhythmic process of respiration and the change in composition between inspired air and expired air, the latter being depleted in oxygen and enriched in carbon dioxide. The analogous sign of global metabolism might be the observation of cyclic fluctuations in the chemical composition of the global atmosphere caused by the activities of the biota.

It was with this idea in mind that I referred to the thirty-year record of carbon dioxide measurements from the Hawaiian island of Mauna Loa (figure 2.1) as an icon of global metabolism. The record clearly shows the fluctuations due to the seasonal cycles of photosynthesis and respiration, as well as a long-term increase attributable to human activities. The uptake of carbon dioxide by plants in the northern summer and its release by respiration in the winter leads to cyclical changes that put atmospheric carbon dioxide at a minimum in September and at a maximum in April—with an amplitude of about 2% of the mean value. Clearly, the seasonal cycles are dominated by the activity of the terrestrial biota of the Northern Hemisphere, which contains most of the global land area.

Global Metabolism: Description, Simile, or Metaphor?

The Mauna Loa records present, in a dramatic and inescapable fashion, one meaning of global metabolism. To speak of global metabolism might be a way of referring to the sum of all biological metabolisms on a global scale. As explored in the previous chapter, there is no dispute that the sum of the metabolic activities of living organisms is of major significance in determining the composition of Earth's atmosphere and hydrosphere. I have already quoted at length from a lecture by a nineteenth-century French scientist, M. J. Dumas, which forthrightly sets out the idea of global metabolism as an aggregation of the activities of the global biota.

Global metabolism, in that sense, is a descriptive term. It may, however, also carry a metaphorical significance. It could suggest that the chemical transformations and translocations of the biogeochemical cycles are analogous to the orderly, self-regulated pattern of reactions occurring within living cells or to the circulation of chemicals from organ to organ within a plant or animal. The key word here is "self-regulated." It is the regulatory properties of metabolic systems that are responsible for physiological homeostasis, for the maintenance of "the constancy of the internal environment."

The concept of the "milieu intérieur" was introduced some thirty years after Dumas's lecture by another French scientist, Claude Bernard. In the second lecture of his "Leçons sur les phénomenes de la vie communs aux animaux et aux végétaux," Bernard wrote the now-famous paragraph:

> The constancy of the internal environment is the condition for free and independent life: the mechanism that makes it possible is that which ensures the maintenance, within the internal environment, of all the conditions necessary for the life of the elements. This enables us to understand that there could be no free and independent life for the simple beings whose constituent elements are in direct contact with the cosmic environment, but that this form of life is on the contrary the exclusive attribute of beings that have arrived at the summit of complication or organic differentiation. (Bernard 1974)

Since Bernard wrote these words, one of the key paradigms of biology has been the distinction between the internal environment, regulated to a very remarkable degree of constancy, and a variable external environment to which organisms may or may not be able to adapt and which may provide the possibility of existence for some species and deny the possibility for others. From this distinction have flowered the disciplines of physiology, with its emphasis on homeostasis, and ecology, with its emphasis on habitat. But if "the constancy of the internal environment is the condition for free and independent life," it is also true that some measure of constancy of the external environment, within the bounds of habitability, is a condition for the existence of any kind of life.

To speak of global metabolism is to raise the question of whether the biogeochemical cycles of the global ecosystem could be self-regulated in similar fashion to the metabolic control of cells and organisms. To speak of global metabolism is to wonder whether such self-regulation could indeed be relevant to the maintenance of global habitability. Are the variables that characterize the global environment restrained within the "life boundary" by mechanisms that are analogous to the mechanisms of physiological homeostasis?

A number of caveats are in order. First, common experience suggests major quantitative differences between a posited global homeostasis and the known levels of organismic homeostasis. For instance, our experience of diurnal, seasonal, and geographical differences in temperature tells us that any global thermostat must be ineffective at this timescale, compared to the exemplary homeostatic system that regulates our own mammalian body temperatures. Second, as we have seen in chapter 1, it is legitimate to question how stable is the global environment. If self-regulatory mechanisms do operate at the planetary level, there have certainly been occasions when their effectiveness was limited. Botkin (1990) has underlined this point:

> We see a landscape that is always in flux, changing over many scales of time and space, changing with individual births and deaths, local disruptions and recoveries, larger scale responses to climate from one glacial age to another, and to the slower alterations of soils, and yet larger variations between glacial ages. (p. 62)
>
> Thus in terms of climate, the cycling of chemical elements, the distribution of species and ecological communities, and the rate of extinction of species, we must reject the possibility of constancy in the

biosphere. If the biosphere has not been in a precise steady-state, then life has not been a precise stabilizing device for the biosphere. (p. 150)

Finally, the analogy of a posited global homeostasis to organismic homeostasis might not be particularly illuminating. Many complex systems may possess some degree of stability conferred by such completely abiotic factors as those dealt with in the previous chapter. (Recall that feedbacks inherent to the relationship among atmospheric carbon dioxide, the weathering rate, and marine carbonate deposition may constitute an abiotic global thermostat.) Stability may be necessary for habitability, but it is by no means sufficient. The environment of a planet that does not support life forms may nonetheless be stable. Such a planetary environment is not necessarily at equilibrium; the internal energy of the planet or the input of solar energy may permit non-equilibrium conditions to exist.

A diagrammatic representation of the circulation of the elements through such a planet's atmosphere, hydrosphere (if it has one), and lithosphere might resemble, at least at first glance, a map of the biochemical pathways of metabolism. The non-equilibrium condition of such a planetary environment would represent a steady-state, stabilized by entirely abiotic feedback mechanisms that maintain the values of its physico-chemical characteristics. The inorganic silicate-carbonate thermostat posited for Earth by Walker, Hays, and Kasting (1981) would typify such a planetary mechanism. Schwartzman et al. (1994) have used this example to raise the general question of whether purely geochemical systems may be capable of self-organization. For such reasons it might therefore be legitimate to speak of "Venusian metabolism" or "Martian metabolism," but the analogy would not be strong. The metaphor of planetary metabolism when used in speaking of Earth is meant to imply more than a superficial similitude.

Two features of the chemical cycles of Earth give added meaning and insightfulness to the metaphor of global metabolism. The meaning is over and above a simple descriptive summation of the activities of the biota on a planetary scale; it is over and above a reference to the resemblance between maps of geochemical cycles and those of biochemical intermediary metabolism. These two features pertain to the roles that organisms play in maintaining the solar-driven steady-state and to the possible involvement of biomolecular feedback mechanisms in determining the characteristics of that steady-state.

First, according to the laws of thermodynamics, any isolated system

Global Metabolism, Geophysiology, and Gaian Metaphors   107

will move toward equilibrium conditions. The maintenance of non-equilibrium conditions thus requires an input of energy. The Earth is not an isolated system, as it receives solar radiation. This solar input provides the energy necessary to maintain even an abiotic Earth at some level of non-equilibrium steady-state. The photosynthetic biota, however, boost the steady-state even further from equilibrium. The transduction apparatus of the biota convert the energy of solar photons into chemical bond energy of a kind and extent unachievable in the absence of life. The steady-state of Earth's environmental chemistry thus does not merely present a superficial analogy to the steady-state of a metabolizing cell or organism. This steady-state is in fact achieved by a balance between, on the one hand, an endergonic production process—photosynthesis, which is brought about by a subset of the global biota—and, on the other hand, exergonic, downhill processes tending to restore equilibrium conditions. The rate of many, though not all, of these latter processes are also determined by biological catalysis. When such catalysis is involved in the constituent processes (and associated feedback loops) of any biogeochemical cycle, then the steady-state levels of the environmental components of that cycle will be influenced by the characteristics of that biochemical catalysis. Three examples of such effects were discussed in some detail in the previous chapter—atmospheric and hydrospheric concentrations of oxygen, carbon dioxide, and dimethyl sulfide. Thus, the first feature of global biogeochemistry that gives appropriateness to the metaphor of global metabolism is that biochemical processes, catalyzed by the global biota, constitute essential components of the feedback mechanisms that maintain a steady-state, far from equilibrium, of many of the constituents of Earth's atmosphere and hydrosphere. Specific characteristics of these biochemical processes may, in fact, determine the steady-state values of some of these constituents.

A second feature of the biogeochemical cycles of Earth that could add depth to the metaphor of a global metabolism is to be found in further details of some of the biologically catalyzed steps of the cycles. Cellular metabolism involves regulation—either regulation of the catalysts of transformation and translocation or regulation of the genes responsible for the production of these catalysts. Later in this chapter I shall review these mechanisms of regulation in more detail. The question that pertains to the worth of the global metabolism metaphor is this: When a process that is responsible at the cellular level for metabolic regulation is also involved in the catalysis of a step in a biogeochemical pathway, can the regulatory features of the internal, biological process also be significant to the chemistry of the external global environment?

Suppose that, as seems likely, the coordination of the global cycles of, say, carbon and nitrogen occurs at the biological level. Does it follow that the molecular mechanisms that coordinate the intracellular metabolic cycles of these elements are also responsible for coordinating the environmental cycles of these same elements?

Consider a hypothetical step in a biogeochemical cycle that is catalyzed enzymically. Suppose the activity of the enzyme responsible for this catalysis is controlled by a feedback loop. If any of the components of such a feedback loop are environmental constituents, then one might justifiably speak of the regulation by the biota of something that might well be called global metabolism.

To sum up the argument to this point: *global metabolism* is an apt descriptive term for directing attention to the influence of biological processes on the physico-chemical characteristics of the environment of the planet Earth. It is also a simile pointing to the resemblance between the control systems stabilizing those global characteristics and those responsible for intracellular and intercellular regulatory mechanisms. Finally, *global metabolism* alludes to the possibility that this latter relationship is not mere simile—that the molecular details of the regulation of biochemical processes may be necessary components of any system analysis at the planetary level.

Gaia and Geophysiology

Such concepts as "global metabolism" and "environmental homeostasis" find their strongest expression in the proposal of James Lovelock (1972) that the planet Earth should be thought of as an organism, "Gaia," in which the atmosphere and hydrosphere are analogous to Bernard's "milieu intérieur" of a metazoan animal. The study of such a system may properly be termed "geophysiology" (Lovelock 1986). Although "global metabolism," "geophysiology," and "Gaia" have somewhat differing connotations, there is considerable overlap in the range of reference of these terms. Thus, I use the plural, "Gaian metaphors" to allude to all three.

The idea of a "living planet" or of a "planetary organism" has aroused considerable criticism, much of which was ventilated at a conference of the American Geophysical Union in 1988 and subsequently published as *Scientists on Gaia* (Schneider and Boston 1991). In the next

Global Metabolism, Geophysiology, and Gaian Metaphors   109

chapter I shall be dealing with the very considerable task of integrating Lovelock's ideas into the mainstream of contemporary biological thinking and, in the final chapter, making some suggestions for solving that problem by pointing toward some molecular biological processes that could subserve a planetary scale geophysiology. For now, let us examine the case for looking at the planet in the new perspective provided by the Gaian, geophysiological standpoint.

Indeed, it was looking at planets that provided the original impetus for Lovelock's putting forward the idea of Gaia. He has recounted his involvement with the NASA planetary exploration program (Lovelock 1988). The concept behind such missions as the Viking probe to Mars called for the design of instrumentation to be carried on remote landing craft that would report back on the detection of the chemical signatures that would indicate the presence of life forms. Lovelock suggested that the information could be obtained with greater certainty by the use of terrestrially based telescopes to analyze spectrophotometrically the chemical composition of planetary atmospheres. The theoretical basis for such an approach was spelled out in a paper entitled "Thermodynamics and the recognition of alien biospheres" (Lovelock 1975). The central point is that on a planet that does not support life the relative abundance of the gases making up its atmosphere will approximate a mixture at chemical equilibrium. Conversely, detection of chemical composition that signifies an atmosphere far from equilibrium would suggest the presence of living systems on that planet. The departure from equilibrium may be evidenced most strongly by gases that are present in the atmosphere as trace components but which, on a sterile planet, would be absent. The hypothesis is illustrated by a comparison of the composition of the atmospheres of Jupiter, Mars, Earth, and Venus (table 4.1). Sagan et al. (1993) recently carried out the appropriate control experiment. They applied Lovelock's criteria to data obtained from a spacecraft fly-by of Earth. "Galileo [the appellation for the spacecraft used] found such profound departures from equilibrium that the presence of life seems the most probable cause."

For our purposes, the next step in the argument is crucial: "The arguments which were used to prove the presence of life on Earth from a mere knowledge of the atmospheric composition can also be used to prove, given the fact that life on Earth is abundant, that the atmosphere is itself almost wholly a biological contrivance" (Lovelock and Lodge 1972). The significance of this move rests in the word "contrivance." The *Shorter Oxford Dictionary* gives, "contrive—plan or design with in-

TABLE 4.1
Composition of the Atmospheres of Venus, Earth, and Mars

	Venus	Earth	Mars
Concentrations (in ppmv):			
N_2	35000	781000	27000
O_2	<30	209000	1300
CO_2	965000	350	953000
CH_4	<4	1.8	<0.02
CO	<35	<0.25	<70
N_2O	<0.1	0.3	—
Pressure (in bars):	90	1	0.0075

genuity or skill"; "contrivance—a thing contrived as a means to an end." It is not the teleological connotation that concerns us here; that problem is taken up in the following chapter. For now, we recognize that the concept of the atmosphere as analogous to "the fur of a mink or the shell of a snail" (Lovelock 1972), or to "the honey or wax of a beehive" (Margulis and Hinkle 1991), implies that the atmosphere is the product of a biological system. The implication becomes stronger when we recall the discussion in chapter 3 that found the facts of geological history to place severe constraints on the variability of atmospheric chemistry. To quote again from Margulis and Hinkle (1991),

> Earth's surface chemistry is aberrant with respect to its reactive gases, its temperature, and its alkalinity. These discordant chemical and physical conditions have been maintained over geologic periods of time. Lovelock's concept, with which we entirely agree, is that the biota (i.e., the sum of all the live organisms at any given time), interacting with the surface materials of the planet, maintains these particular anomalies of temperature, chemical composition, and alkalinity.

Two ideas are involved here. First is the posited existence of "an adaptive control system that can maintain the Earth in homeostasis" (Lovelock 1990). Second is the suggestion that the ability to achieve such control is "a collective property of the growth, activities, and death of the myriad of populations that comprise the biota" (Margulis and Hinkle 1991). It is not easy to distinguish the contributions of these two ideas to the overall picture of Gaia, but it is important to recognize that they must *both* be present. If the idea of a control system is allowed to predominate, then "Gaia" becomes merely a pet name for the system, much as in the early days of computers people spoke of "ENIAC" or "UNIVAC." If, on the other hand, the idea of a control system is not given sufficient prominence, then "Gaia" becomes a fanciful synonym for the global biota. The idea of Gaia gains its potency by identifying the global biota as the system responsible for maintaining the anomalous non-equilibrium properties of Earth's atmosphere and hydrosphere in a steady-state.

Justifying the Gaian Metaphors

Much ink has been spilled on the question of whether it is appropriate to speak of the Gaian planet as a living organism. The issue is not entirely a semantic one. The idea of the collective activity of the global biota does raise important challenges for evolutionary theory—certainly Margulis believes it to do so: "That the Gaia concept cannot be framed by the stilted terminology of neo-Darwinistic population biology is not surprising since Gaia is a hypothesis based in sciences that neo-Darwinism proudly ignores" (Margulis and Hinkle 1991). I take this to mean that biologists, attempting to construct a theoretical framework within which the history of life on this planet is to be understood, have ignored the physics and chemistry of the planetary environment. The biota are considered in isolation and not as part of a planetary system.

But the other pole of the Gaian idea constitutes a corresponding challenge to geochemists—for while they have been willing to see the Earth's crust, hydrosphere, and atmosphere as constituting a more or less tightly coupled system of cyclic processes, they have been relatively unwilling to assign to the biota the dominant role required by the Gaian scenario. Three quotations from geochemists participating in the 1988 American Geophysical Union symposium are indicative of geochemical caution in this regard:

> The presence of abundant oxygen in the atmosphere of the Earth, as compared with other planets, is an indirect indication of the presence of life.... Nevertheless, there are major controls on the level of atmospheric oxygen that are geological in origin and not dependent on life processes. (Berner 1991)
>
> It seems likely, then, that the CO_2 content of the atmosphere is influenced by the biosphere, but that the biosphere is a poor regulator of pCO_2. To that extent CO_2 is not a good Gaian gas ... the relative

constancy of pO_2 during the past 300 million years owes as much to inorganic chemical processes and to physical oceanography as it does to the biosphere. (Holland 1991)

Summing up the performance of life processes in the global carbon cycle, it seems justified to state that the Earth's biota stars at a very prominent level, but certainly does not run the whole show. (Schidlowski 1991)

It is perhaps irrelevant whether or not we call the global system "Gaia." Lovelock (1979) concedes that one might use some such "barbarous acronym" as "BUST" or "BUSH" (Biocybernetic Universal System Tendency/Homeostasis), though the identification with the earth mother goddess of Greek mythology has contributed both to a favorable popular reception and to scientific resistance. The classification of the global control system as "a living organism" (Lovelock 1988:40) or as "quasi-living" (Lovelock 1991b:31) has exacerbated the confrontation with biological theory, which is the topic of the next chapter. The term *geophysiology* is perhaps only slightly less offensive to many biologists than is the direct organismic metaphor, as at least since the eighteenth century the word *physiology* has been used only for living things. Lovelock (1986) uses *geophysiology* to denote a systems approach to the interactions between the global biota and their physico-chemical substrates that give rise to the anomalous characteristics of Earth's surface among the planets of the solar system. The third Gaian metaphor, "global metabolism," permits the widest range of meaning from, at one extreme, a weak recognition of the involvement of the metabolism of the biota in geochemical processes to "Gaia-as-organism" hypotheses at the strong end of the spectrum. Between these two positions one may ask a whole range of questions about the orderliness of biogeochemical cycles.

Perhaps two of these questions may be most readily put in the terminology introduced in the previous chapter. First, of the factors listed as responsible for restraining global environmental variables within the life boundary, how important is "negative feedback" in comparison to "luck," "size," and "chemical equilibria"? If feedback loops constitute a major factor in planetary stability (a necessary condition for habitability), then we are justified in speaking of a global control system. Second, do the global biota play a major role in those feedback loops? If so, then we would be justified in speaking of a global biological control system.

In chapter 3 I used the examples of oxygen, carbon dioxide, and dimethyl sulfide as examples of global feedback systems in which the biota play major roles. There may be many others. One of the major tasks of geophysiology would be the identification of the feedback loops that achieve global regulation. At a time when the stability of the global environment is of such deep and widespread concern, it is important to identify all possible factors in global stability. It is particularly important to identify possible feedback mechanisms because it is by no means obvious that a stabilizing mechanism will itself be robust. To use a simple analogy, the amount of force required to vandalize a domestic thermostat is far less than that required to wreck the furnace. The recognition and amplification systems responsible for conferring a remarkable degree of stability on the anomalous physico-chemical characteristics of Earth may themselves be relatively susceptible to perturbation by human activity. We do not know. Indeed, it is not clear how important such mechanisms are, but in a geophysiological perspective one must accept the possibility that anthropogenic pressure on some seemingly insignificant component of the global ecosystem could have widespread effects on global habitability for an unknown range of species. The determination of whether such mechanisms exist and their identification and characterization is a major issue for global environmental management.

What are the features that distinguish the biological mechanisms of metabolic control from the inorganic feedback processes that might be involved in determining the physico-chemical characteristics of the surface of a sterile planet such as Venus or Mars? The distinction between geochemical feedback and the regulation of metabolism arises from the differences between, on the one hand, systems in which the rate of flow through a reaction sequence is governed by simple mass action principles and, on the other hand, systems in which individual steps in the sequence are catalyzed by biogenic enzymes, the activity of which may be modulated. Before turning to that comparison, it is important to emphasize that the modulation of enzyme activity does not evade the law of mass action. Differences of behavior between abiotic and biotic control systems are attributable to the propensity of the catalysts in the second type to be modulated in a highly specific manner.

Any such propensity of a catalytically active protein has been selected in evolutionary history. Because that history varies for every species, particular mechanisms of catalytic modulation are not predictable. Consequently, each system of the second, biochemical, type will have its own characteristics. For example, the control system imposed on the breakdown of glucose in the muscle of bumble bees (*Bombus* sp.) differs from that found in very similar bees (*Psithyrus* sp.). The differences can be correlated to physiological and behavioral differences (Newsholme and Crabtree 1976). Even within a given animal, the control of glucose breakdown in liver differs from the controls imposed in excitable tissues such as muscle and nerve.

Nonetheless, it has proved possible to detect patterns of behavior among biochemical systems. Thus a major question for global metabolism is whether such patterns can be discerned in the biogeochemical cycles. A brief review of these patterns follows. (Readers who already possess a good deal of familiarity with biochemistry may wish to skip to the next chapter.)

Patterns of Metabolic Control: (1) Feedback Inhibition

One of the most typical patterns of modulation found in biochemical systems is that of *feedback inhibition*. Consider the two schemes of figure 4.1. The scheme of figure 4.1a is a simple sequence of consecutive chemical reactions. In the scheme of figure 4.1b it is supposed that the enzyme catalyzing the reaction A → B is inhibited by D, a state of affairs known as "end-product inhibition." Detailed mathematical comparisons of the behavior of such systems are available, but the principal advantages of the scheme of figure 4.1b are intuitively obvious. The system exhibiting feedback inhibition will have lower levels of intermediates B, C, and D. The steady-state level of the product, E, will be more stable—both with respect to changes in the concentration of substrate, A, and with respect to demand. Particularly with respect to demand, if an alternative source of the product, E, becomes available, the pathway from A will be shut down.

Savageau (1976) has discussed the time-dependent behavior of systems where the end-product is an inhibitor of an early step in its biosynthetic pathway and has attempted to elucidate optimal design principles for such systems. The two most important factors are the strength of product inhibition and the effective path length. The strength of product inhibition is equated to the apparent order of the first reaction, A → B, with respect to the product E (≤ 0, because E is an inhibitor). The sensitivity of the steady-state to changes in substrate supply or product demand decreases as the numerical value of the apparent kinetic order increases. But above a certain threshold value, the time-dependent

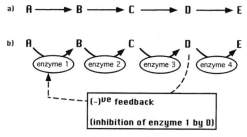

FIGURE 4.1 (a) A simple sequence of consecutive reactions. (b) A sequence of enzymically catalyzed reactions with negative feedback.

response to perturbation exhibits oscillatory instability. The onset of unstable transient behavior is also affected by the length of the pathway from the inhibited step to the end-product. For a given strength of product inhibition, pathways in which the number of intervening slow reactions is above some threshold value will be dynamically unstable and perturbation will lead to oscillations of increasing amplitude. The effective length of the pathway is not defined by the number of intermediate steps, only by the number of steps that are kinetically relevant. To quote Savageau (1976):

> Consider the situation in which the corresponding kinetic parameters of the reactions in a sequence are sufficiently different in value that some of the reactions occur at a much faster rate. Since these are irrelevant to the system's behavior, which is determined by the slower (rate-limiting) reactions, the pathway may be considered from a kinetic point of view to have fewer reactions than it actually does.

Even the definition of the "end-product," which may seem fairly obvious from figure 4.1, does in fact require some thought. For example, the end-product of glycolysis in animal tissues might appear to be lactate or acetyl-coenzyme A. But from the point of view of biological energy requirements, it might be better to think of adenosine triphosphate or even fatty acids as the end-product of glycolysis. Indeed, the network of metabolic reactions is such that the biologically appropriate modulator may not be an end-product in the literal sense at all. A com-

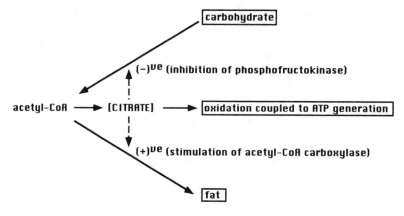

FIGURE 4.2 Regulatory effects of a metabolic intermediate, citrate.

pound such as citrate—which plays a part in a number of physiologically significant reaction sequences—may act as a signal, integrating the flow through these pathways (figure 4.2).

If the definition of "end-product" is contextually dependent (and the same argument applies to the definition of "substrate"), then the definition of the length of a metabolic pathway is equally open. The number of enzymically catalyzed steps in defined metabolic sequences will be highly variable. However, as already pointed out, not all of these steps will have regulatory or integrative significance. Only those steps that are rate-limiting can be significant in this regard.

The idea of rate-limiting reactions is of great importance in understanding the control of metabolism. A number of diagnostic features are characteristic of the controlled, rate-limiting steps in a reaction sequence. (1) Such reactions are typically irreversible, i.e., they are associated with a large negative change in free energy. (2) The steady-state ratio of products to reactants ("mass action ratio") is far from the equilibrium value. (3) The relationship between the velocity of the reaction and substrate concentration is not hyperbolic but sigmoidal. (4) The activity of the enzyme catalyzing such a reaction is markedly affected by modulators (low molecular weight compounds that specifically bind to the enzyme and change the activity of its catalytic site). (5) Alternatively to (4), the enzyme may be subject to reversible covalent modification, e.g., phosphorylation and dephosphorylation of side chains of serine residues, and the activity of the enzyme may be governed by the extent of modification—the modification being itself catalyzed by other enzymes, e.g., protein kinases and phosphatases. These last three characteristics of controlled, rate-limiting steps in reactions are peculiar to biochemical feedback mechanisms and thus distinguish them from the geochemical mechanisms discussed in chapter 3. If the persistence, on a geological timescale, of habitability is a function of geochemical stability, then it must also be recognized that, within the life boundary, adaptation to the environmental constraints of Earth's environment has been made possible only by virtue of molecular mechanisms that ensure biochemical stability. The relationship of molecular regulatory mechanisms to biological adaptation is taken up in the next chapter.

Patterns of Metabolic Control:
(2) Allosteric Modulation

Prominent among the biological mechanisms of enzyme regulation is allosteric modulation. *Allosteric modulation* is the modification of the activity of an enzyme's catalytic site by the binding of a ligand to another, allosteric, site on the protein. The model for all such mechanisms is hemoglobin. Hemoglobin is not itself an enzyme, but, in its ability to bind oxygen in the lungs or gills and deliver such to the respiring tissues, it has much in common with the enzymes. The binding curve relating percentage saturation to partial pressure of oxygen (pO_2) is sigmoidal (S-shaped). As a consequence, there is a range of oxygen concentrations in which small changes in that concentration bring about large changes in the ratio of oxyhemoglobin to deoxyhemoglobin. Physiologically, this effect is very important for the supply of oxygen to the respiring tissues and the uptake of oxygen in the lungs or gills. The general significance of such a relationship in enzyme regulation is that, for allosteric enzymes, there is a range of substrate concentrations in which the order of dependency of reaction rate on substrate concentration is greater than 1. For higher-order relationships the kinetic behavior approaches an on-off switch.

The molecular basis of such complex behavior is now well understood. In the case of hemoglobin, each molecule contains four binding sites for O_2, each binding site being located on one of the four subunits (two α and two β chains) that make up the tetrameric molecule. The binding of O_2 by one of the sites brings about an increase in the affinity for O_2 of the unoccupied sites. Because the sites are too far apart on the hemoglobin molecule to interact directly, the effects must be mediated by changes in the molecular structure. These structural changes have now been mapped in considerable detail (Perutz 1989).

Such effects of ligand binding on the affinity of unoccupied sites for the same ligand are known as *homotropic effects*. The effects of the binding of a ligand on the affinity of binding sites for other ligands are termed *heterotropic effects*. Hemoglobin exhibits both types. The best known heterotropic effects on the binding of O_2 by hemoglobin are brought about by the binding of protons (H^+) and specific organic phosphate esters.

Patterns of Metabolic Control: (3) Enzymic Cascades

Allosteric modulation is not the only means of regulating the activity of enzymes. Another important mechanism is the enzymically catalyzed reversible modification of specific amino acid residues. Because the modification is brought about by an enzyme, which may itself be similarly activated, a cascade effect results.

Each molecule of activating enzyme may transform many molecules of inactive enzyme to the active state; each molecule of enzyme thus activated can transform many molecules of substrate. (The enzymes involved may be subject to allosteric control.) Any such cascade must be initiated in some manner. For instance, the first such cascade to be recognized and studied in detail is the response of phosphorylase (an enzyme that plays a crucial role in carbohydrate metabolism) to hormonal stimulation (figure 4.3). The initiation of the cascade is brought about by the intracellular "second messenger," cyclic AMP, which activates the first catalyst of the cascade. Cyclic AMP is produced in the cell by an enzyme of the cell surface that is responsive to external signals in the form of ligands that bind highly specifically to sites on the cell surface. The binding of these ligands activates the enzyme responsible for the production of cyclic AMP.

The details of these powerful control systems are not relevant to the understanding of biochemical adaptation necessary here. What is essential to the argument that follows is the recognition that these mechanisms, the allosteric mechanisms and the enzyme cascade mechanisms both depend crucially upon the ability of proteins to bind low molecular weight ligands in a highly selective and specific fashion.

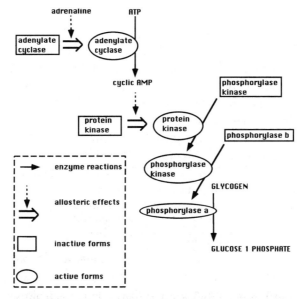

FIGURE 4.3 The archetype of enzymic cascades—the response of phosphorylase to hormonal stimulation. Note that the amplification at each stage means that a single molecule of adrenalin could effect a major "switch on" of glycogen breakdown.

Patterns of Metabolic Control: (4) Gene Regulation

Both the allosteric mechanisms and the cascade mechanisms regulate cellular metabolism by altering the catalytic activity of protein molecules, enzymes, already present in the cell. Another way in which metabolism can be regulated is by control of the number of enzyme molecules present.

Enzymes, like all proteins, are synthesized by the linear assembly of amino acids to form a polypeptide chain. The sequence of amino acids in each molecule corresponds to a sequence of nucleotides in a polyribonucleotide molecule, messenger RNA. The translation of the nucleotide sequence into an amino acid sequence involves activation of the amino acids by formation of amino-acyl t-RNAs, alignment of the anticodons of the amino-acyl t-RNAs with the codons of the messenger RNA, and the sequential formation of peptide bonds. The messenger RNA itself is modified (e.g., by removal of noncoding sequences:

introns) from a polyribonucleotide molecule produced by transcription of a given portion of the chromosomal DNA.

The best understood of the mechanisms for regulating the amount of any given enzyme is the control in procaryotes of the transcription of the particular stretch of DNA that codes for that enzyme. In the operon model (Jacob and Monod 1961) the initiation of transcription is blocked by the binding of a repressor protein to specific DNA sequences. In the presence of an inducer (physiologically, the substrate of the enzyme in question) the repressor protein dissociates and transcription is initiated. The RNA transcript is then translated to give catalytically active protein. Conversely, in control of gene activity by product, the binding of product to the repressor protein gives a complex that can block transcription of those stretches of DNA responsible for the synthesis of the enzyme leading to the product. As in the case of activity control, the details of this process are not germane to the understanding of biochemical adaptation needed for our purposes. Note, however, that the dissociation of the repressor protein from the DNA follows upon the binding by the repressor of a ligand of low molecular weight. The repressor protein possesses dual binding specificity, which is the ability to "recognize" both a specific nucleotide sequence in DNA and also the ligand that activates substrate derepression or product repression (cf. Zhang et al. 1987). Once again, as in the case of the allosteric mechanisms and cascade mechanisms for the control of enzyme *activity*, the genetic control of enzyme *levels* also depends crucially upon the selective and specific binding capability of particular proteins.

Organismic Homeostasis

The ability of proteins to bind other molecules in a highly specific fashion is not only important for the control of intracellular metabolism. It is also important for the metabolism of multicellular organisms. Constancy of composition of the "milieu intérieur" in the face of fluctuations in the external environment is assured by feedback loops in which changes in the chemistry of the bodily fluids are sensed by a neuronal or endocrine cell. Activity of the cell in response to the chemical change (for example, the production of insulin by the β-cells of the pancreatic islets in response to an increase in plasma glucose) will result in changes in the metabolic activities of other cells such as to counteract the change (in the example given, the increased transport of glucose across the membranes of muscle cells and a decrease in gluconeogenesis by liver

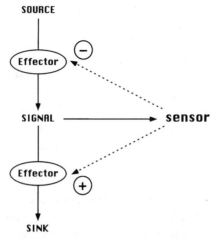

FIGURE 4.4 A generalized feedback loop. Note that if the concentration of the signal rises, its rate of production is slowed and its rate of transformation accelerated.

cells). The responsiveness of the neuronal or endocrine cell to a specific chemical signal and the responsiveness of the target cells to a transmitter or hormonal signal are all further examples of the way in which metabolic control depends upon chemical recognition of ligands by proteins. The sensing cell must possess proteins that can selectively bind that chemical species, the concentration of which is being monitored; the target cell must possess proteins that can selectively bind the transmitter or hormone that is being released by the sensing cell.

It is possible to construct a general diagram for all these regulatory mechanisms, as in figure 4.4. The interaction of the sensor with the effector varies considerably among the examples already described. In the allosteric mechanism of enzyme regulation, the sensor and effector are contained within the same oligomeric protein molecule, although they may reside on separate regulatory and catalytic subunits. Transmission is intramolecular, involving adjustments in the three-dimensional structure of the enzyme. In the cascade mechanism, the transmission is intracellular but intermolecular, involving the enzymically catalyzed modification of the effector. In the genetic (operon) mechanism, the transmission from sensor (repressor) to effector (gene product) is again within the cell, but the interaction between molecules is indirect and involves the cellular machinery of transcription and translation. In the final example, metabolic regulation at the organismic level, transmission takes place between cells through the extracellular environment.

By extrapolation, to speak of metabolism at the global level would seem to suggest transmission among organisms through the external environment. This may be a weakness in the analogy of global biogeochemical cycles to biochemical metabolism, because organism-to-organism interactions seem likely to be weaker than interactions within the one organism. A more damaging weakness shows up when we look at what lies behind the ability of the sensor to recognize the appropriate signal.

Although there is considerable variation in the details of the interaction of the sensor with the effector, the factor common to all the cases briefly outlined in this chapter lies in the mode of interaction of the sensor with the signal. The sensor must *recognize* the signal. The recognition consists of the binding of the signal molecule by a protein. This binding has to be highly specific. Metabolism is characterized by an orderliness that distinguishes it from a set of reaction sequences interacting in a straightforward physico-chemical fashion, and this orderliness depends upon the specificity of signal recognition, i.e., upon the specificity of the interaction of signal molecules with sensor proteins. This signal recognition (the ability to bind a ligand of highly defined chemical structure) is, in turn, dependent upon the three-dimensional structure of the proteins that reversibly bind and release the signal ligand.

The mutational origins of these highly specific three-dimensional structures characteristic of the biological realm and the selection of those that enhance the fitness of the individual organisms that possess them constitute the molecular foundation of modern evolutionary theory. In the next chapter I turn to the difficulties of integrating the Gaian metaphors of orderliness in global biogeochemistry with this account of the genesis of metabolic orderliness within organisms.

5
Teleonomy and the Biological Critique of Gaia

"Why then has Gaia been so unpopular among scientists?" James Lovelock raised (and offered an answer to) this question in his 1990 defense of the Gaia concept published in *Nature*. It is perhaps unfortunate that the question is posed in terms of a popularity contest. And Lovelock's own attempt at an answer is couched in terms of the sociology of scientific controversy. I would like to take a more detached standpoint by asking, What is it about the Gaia hypothesis that makes the concept difficult to integrate into the mainstream of contemporary biological thought?

Before tackling the question directly, it is necessary to look at the topics of chapters 1 and 3 from a biological point of view. We started with the question of habitability. After a brief excursion into the anthropogenic factors altering the chemical and thermal regime of the planet (chapter 2), I introduced in chapter 3 the concept of a "life boundary," which had first been put forward in 1966 by Peter Weyl. In his definition of the life boundary Weyl was careful to point out that "because of evolutionary adaptability, the life boundary is not necessarily static." We now need to examine this proviso in some more detail. Afterward we shall be better able to address what it is that underlies the lack of enthusiasm for (and outright hostility toward) the Gaia concept among many biologists.

Habitability and Adaptation

Habitability cannot be defined by any set of environmental characteristics without reference to the adaptive capabilities of the biota. The con-

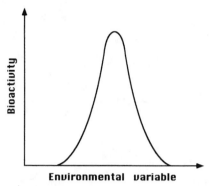

FIGURE 5.1 The bell-shaped response to a varying environment. This particular example shows an optimum at the center of the biologically relevant range of values of the environmental variable. But asymmetrical curves would be more common.

cept of a life boundary lays stress on the limits of biological adaptability. Indeed, the life boundary may be thought of as defining the limiting environmental conditions beyond which adaptation cannot occur. It is, however, important to realize that the life boundary cannot be conceptualized on purely *geo*chemical grounds—it cannot be defined without reference to *bio*chemistry.

Looked at from a biological point of view, the location of the life boundary lacks both precision and accuracy. It lacks precision because there can be no sharp edge to the range of conditions within which some living organisms may be able to survive and reproduce. Many biological activities are related to environmental parameters in a bimodal fashion. That is to say, a plot of the activity against the value of the environmental parameter will be a bell-shaped curve, with an optimum and with "tails" at the extreme, as in figure 5.1. The same biological activity may possess a similarly shaped relationship to some other environmental variable; so, somewhat naively, one could portray the relationship of the activity to both variables as in figure 5.2. Here, a life boundary is represented on the bottom (x-y) plane. In the case depicted, the life boundary is circular, with the most favorable conditions for biomass growth being at the center of the circle. Now consider that, in the real world, the activity represented on the z-axis would most likely

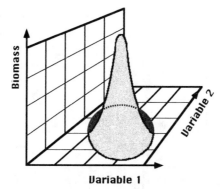

FIGURE 5.2 A possible range of biological responses to two environmental variables. In an aquatic system, variable 1 might be salinity, variable 2 might be temperature. The circle depicted on the x-y plane constitutes a "life boundary" (compare figure 3.2). At combinations of temperature and salinity outside the circle no growth occurs.

approach zero asymptotically. Thus the exact edges of the circular life boundary cannot be precisely located. Then too, the three-dimensional shape of the activity-environment relationship depicted in figure 5.2 could be highly variable. Depending on the nature of the activity and the environmental variables, the sides could be more or less steep, the peak could be substantially off-center and more or less flat-topped. More important, it is highly improbable that a two-variable life boundary would take the cylindrical form that might be implied by a life boundary defined purely by geochemical considerations.

Only if habitability is thought of as an all-or-none property would a habitability-environment relationship, represented as in figure 5.2, be in the form of a cylinder. But habitability is not an all-or-none property. It is indeed difficult to provide a quantitative measure of habitability, but it is not illogical to speak of the Antarctic as being less habitable than Amazonia. Biomass is some sort of a measure of habitability, and it is with that in mind that I labeled as "biomass" the z-axis of figure 5.2. Environmental space is multidimensional rather than three-dimensional. But insofar as the life boundary is constituted from curves in which biological activity approaches zero asymptotically, it will remain true that the limits of such activity cannot be precisely defined.

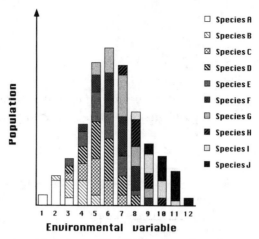

FIGURE 5.3 A possible distribution of species as a function of some environmental variable. Species A and J do best near (but not at) the extreme conditions. Note that the curve defined by this bar chart resembles the bell-shaped curve of figure 5.1, which aggregates all species and ignores specialization and biodiversity.

The representation of figure 5.2 also helps us to understand why the location of the life boundary not only must lack precision but also lacks accuracy. It lacks accuracy because there is no way of determining what the limits of biological adaptation are. If one were to think of the z-axis as biomass, then that biomass would be made up of the biomass contribution of many species, and those species would not be uniformly distributed inside the life boundary. In relation to one of the environmental variables, the distribution might resemble that of figure 5.3. Typically, as the life boundary is approached, the number of species is reduced, as is their stable population. Total biomass is smaller and population structure is simpler as the environment becomes harsher. However, figure 5.3 also demonstrates that, for some species, these harsher conditions actually provide an ecological niche. It is not necessarily true that environmental conditions close to the life boundary represent optimal conditions for survival and reproduction of these particular species; it merely has to be the case that they are adapted to survive and reproduce under circumstances that are outside the environmental range of other species (either competitors or predators) which lack the appropriate adaptive mechanisms. From the biological side, it is these outlier species that set the limits of habitability at any particular point in time.

Adaptation in Molecular Terms

What is the biochemical basis for the life boundary? What limits the environmental range within which the patterns of cellular metabolism can be maintained? The answer to these questions lies in the stability of the macromolecules that make possible these regulated cycles. The ordered reactions and translocations that make up the metabolic cycles within cells are catalyzed by macromolecules, principally by proteins. This catalysis is brought about as a consequence of the highly specific binding of the substrate to a well-defined site on the protein macromolecule.

At the end of chapter 4 I made reference to the dependence of signal recognition upon the three-dimensional structure of the biological receptor. We now have to look at this dependence in more detail. Enzymologists recognized a long time ago that the specificity of enzymes with respect to their substrates suggested that there must be a precise stereochemical complementarity between the catalytic site of the protein and the substrate. In the last thirty years or so this explanation of specificity has received ample direct confirmation from the determination by X-ray diffraction of the structures of proteins and protein-ligand complexes in crystals. And it is now possible to identify which amino acid residues of the polypeptide backbone interact with the ligand. Comparison of the structures of the protein and of the ligand separately with that of the protein-ligand complex reveals the extent to which the conformations of both the protein and the ligand are changed in forming the complex (Phillips 1966).

Such studies leave no doubt that the binding specificity of enzymes is attributable to their three-dimensional structure. It is the twisting and bending of the polypeptide chain that places particular charged or hydrogen bonding residues into the precise topological relationship that is necessary for interaction with functional groups on the ligand. This twisting or bending may also create crevices lined with hydrophobic residues, which may accommodate hydrophobic groups on the ligand and thus minimize the energy of interaction with the aqueous micro-environment. Such interactions provide the molecular basis for the affinity and specificity of components of the cellular translocases, such as the phosphate transporter (Luecke and Quiocho 1990) which, as described in chapter 3, may play a key role in determining salient features of the global environment. The binding of ligands by proteins is accompanied by a decrease in free energy equal to $-RT\ln K_a$, where K_a is the equilibrium constant for the association of ligand with protein. This

K_a is typically large, between 10^3 and 10^6. The Michaelis constant, K_M, of enzymology is related to the dissociation constant, $1/K_a$. A high value of K_a (a high affinity, a low value of K_M) means that maximal reaction velocity is reached at low substrate concentration. But if K_a were too large, the association of ligand would be irreversible, and the biological function of the transitory binding would be lost.

Thus, both the specificity of binding and the efficacy of catalysis depend upon a close complementarity between the shape of the substrate molecule and that of the active site of the enzyme. This latter is a feature of the overall three-dimensional structure of the protein; perturbation of that structure may distort the active site and render it catalytically inactive. How does a protein molecule acquire this highly specific three-dimensional structure essential for its biological function? It has been a commonplace of molecular biology since the experiments of Anfinsen (1962) on ribonuclease that the three-dimensional structure of a protein molecule—which both confers on the molecule its specificity of binding small ligands and sets the equilibrium of their association/dissociation—is completely determined by the amino acid sequence of that molecule. This statement is accurate in the sense that the transition from the linear product of translation to the correctly folded protein requires no further input of genetic information. However, the three-dimensional structure is environmentally sensitive. It may thus be proper to think of the three-dimensional structure of a protein as a phenotype, while the linear amino acid sequence is a direct translation of the genotype.

Much effort is currently being spent on the elucidation of the rules that relate primary structure (amino acid sequence) to secondary structure (helices, sheets), to tertiary structure (bends and folds), and to the way in which the folded subunits associate in oligomeric proteins. All this is, of course, specific to a given micro-environment, which may include other proteins—the "molecular chaperones" (Ellis and Hemmingen 1989; Martin et al. 1991; Todd et al. 1994). This quest for the rules governing the transformation of two-dimensional to three-dimensional structures, which has been referred to as the breaking of the second half of the genetic code (Goldberg 1985; Kolata 1986), is fraught with matters of considerable thermodynamic complexity. The number of possible conformations for a long polypeptide chain is extremely large. Although many of these structural possibilities may be ruled out on steric grounds, a large number of permissible states remain. The task of determining energy minima and choosing among these minima is formidable. Fortunately, an increasingly large number of structures are now becoming known from X-ray crystallographic studies, while nuclear magnetic resonance spectroscopy holds out the promise of structure determination in solution. Thus certain empirical generalizations are becoming available to assist in the prediction of three-dimensional structure from a known amino acid sequence (Bowie et al. 1990). As the understanding of the detailed stereochemistry of these delicate and elaborate structures becomes more complete, so will our understanding of the environmental boundary within which these structures are stable and remain functional.

Even the current incomplete state of knowledge suffices for our purposes here. We can indeed state the problem of adaptive mechanisms in molecular terms, as follows. The range of habitable environments for organisms may be extended in one of two ways: either the molecular machinery of the cells is modified so that it is able to function in the new environment or some way is found to shield the molecular machinery from the adverse effects of the new environment. It is important to emphasize that both strategies depend in a fundamental way on evolutionary changes in protein structure. Hochachka and Somero (1984), in their discussion of the ways in which organisms adapt to highly saline environments, put the matter this way: "The use of compatible solutes and counteracting solute systems shifts the evolutionary changes entailed in osmotic regulation from proteins in general to the specific proteins involved in regulating organic osmolyte concentrations."

The important point to be made here is not the contrast between the two modes but that both modes of adaptation depend upon the evolutionary selection of new proteins. Put in genetic terms, adaptation means acquiring a genotype which, in interaction with the new environment, generates the phenotypic characteristic of being able to maintain cellular function. Put in biochemical terms, adaptation by the modification of the cellular machinery means acquiring the ability to synthesize new proteins capable of function in the new environment. In contrast, adaptation by conservation of the micro-environment of an organism's proteins means acquiring the ability to synthesize new proteins capable of effecting the necessary regulation of the micro-environment, so that the old cellular machinery can continue to function. Organisms capable of surviving in a hostile external environment may do so by maintaining a very different internal micro-environment that permits biomolecular function. This ability (e.g., transmembrane "pumping") will be conferred by the possession of specific proteins (e.g., transmembrane "pumps").

Biochemical adaptation, therefore, is achieved by two distinct strategies. One buffers the micro-environment of the cellular machinery (thought of as an extension to the cellular level of Bernard's "milieu intérieur"). The other changes the molecular characteristics of that machinery so as to keep it functional in a less favorable environment. It is the former that would appear to be of greater relevance to the metaphor of global metabolism. This metaphor casts light on the stability of the global environment rather than on the limits of adaptation.

Is it permissible to draw an analogy between the adaptive achievement of a micro-environment that permits biomolecular function and the achievement of a global environment that is habitable? The two previous chapters have outlined some of the processes that serve to buffer the global environment and the role the biota play in these processes. There are, however, important differences between the processes stabilizing the external macro-environment and the mechanisms that regulate the intercellular and intracellular micro-environments. A major clue to the differences lies in the genetic origin of the latter. Recall the argument of chapter 4 that, whereas the processes that underlie geochemical stability have their origin in broadly applicable physico-chemical principles such as the law of mass action, the mechanisms underlying physiological homeostasis arise by specific mutation of the genetic material and are peculiar to those genotypes.

The Evolution of Physiological Regulation

Biological function depends upon three-dimensional structures that arise from the linear sequence of amino acid residues arrived at by transcription and translation of specific genes. Such a broad statement, providing the link between biochemical adaptation and natural selection, would command general assent. However, making the linkage in specific cases from genotype to phenotype and from phenotype to Darwinian fitness of individuals is not an easy task.

Feder (1987) has pointed out that it is not obvious what the relationship will be between physiological variation and fitness. Figure 5.4 shows two extremes of a wide range of possible relationships. The physiological characteristic measured on the abscissa in both upper and lower graphs is a quantitative phenotypic character that can affect the reproductive success of the individuals possessing that character. In the hypothetical case of figure 5.4a, the phenotype must be closely matched to the environment; any significant variation in the phenotypic charac-

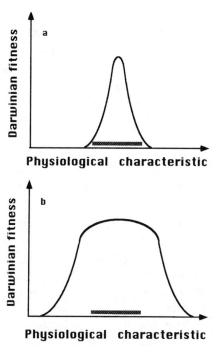

FIGURE 5.4 Two possible ways in which Darwinian fitness might be related to the range of some given physiological characteristic. *Top:* In this hypothetical case, the phenotype must be closely matched to the environment; any significant variation in the phenotypic character will be strongly selected against. *Bottom:* In this case, variation of the phenotype within the limits of the horizontal bar will have little effect on fitness.
SOURCE: After figure 3.4 of Feder 1987.

ter will be strongly selected against. On the other hand, in the case depicted in figure 5.4b, variation of the phenotype within the limits of the horizontal bar will have little effect on fitness.

If there is no simple relationship between phenotypic physiological characters and fitness, it is equally difficult to derive physiological characteristics from knowledge of the genotype. Koehn (1987) discusses the evidence for the genetic component in the variability of physiological characteristics. He points out:

Virtually all physiological characteristics are highly variable: physiological variation has largely been attributed to environmental influences that act upon complex metabolic systems (i.e., nongenetic), or at best to polygenes. Since physiological traits result from a complex metabolism, it has been reasonably assumed that a significant proportion of the variation in any specific physiological trait could not be caused by variation of one, or a few, specific genes.

Despite these difficulties, a start is being made on the correlation of specific genes with physiological function and with survivorship (Powers 1987). For specific enzymes that are found as different isozymes, corresponding to different allelic forms of the gene coding for the enzyme, it has been observed that the different isozymes possess different geographic distribution. These allele frequency clines can be correlated with changes in the values of environmental variables, such as temperature or salinity. Further, the differing catalytic properties of the isozymes can be shown to give rise to changes in cellular metabolism (DiMichele et al. 1991) that are adaptive with respect to the environmental gradient. Finally, these differences in metabolism are responsible for differences in performance and in the response to experimental manipulation of environmental parameters. A causal chain is thus indicated from gene to enzyme to physiological function to Darwinian fitness. Such studies, and related investigations of the inheritance of quantitative physiological traits (Arnold 1987), provide some empirical basis for the assertion that evolutionary adaptation occurs by selection of genes that confer the ability to synthesize novel proteins. It is also noteworthy that, in some of the cases studied, the advantage conferred by possession of a particular allele seems to occur because the gene product, a specific isozyme variant, brings about a more favorable microenvironment for other proteins.

If metabolic regulation depends upon signal recognition by a protein sensor, and if this recognition is dependent upon a three-dimensional structure that is genetically determined, then it is possible to consider the evolutionary genetics of metabolic control. Selection pressures might be expected to favor an appropriately regulated system over an otherwise similar system that is uncontrolled. Once more, the allosteric interactions in the hemoglobin molecule provide a model. The cooperative effects of ligand binding are not the same in all species. It is therefore possible to ask whether the differences may be adaptive. Hochachka and Somero (1984) discuss the physiological significance of the differing functional behavior of fish hemoglobins. They interpret the possession of these hemoglobin variants in terms of the ecological physiology of these species. The ecologically significant variations in the effects of ligand binding are known to arise from relatively minor changes in the amino acid sequences of the proteins.

The comparative study of regulatory enzymes is at a much less advanced stage than the study of hemoglobin variants, but there is no reason to doubt that the lessons from hemoglobin will be generally applicable. It is likely that there will be interspecies differences in regulatory behavior that can be interpreted in terms of physiological adaptation. These differences in the regulatory properties of the enzymes will be attributable to changes in protein structure, sometimes involving relatively few amino acid substitutions in the protein backbone.

The adaptive significance of perhaps slight genetic differences in enzymes is also well illustrated by certain diseases in humans that are known to be inborn errors of metabolism. For instance, some forms of hypercholesterolemia arise because of a loss of the end-product inhibition (figure 4.1b) of HMG-CoA reductase, a key regulatory enzyme in the biosynthetic pathway for cholesterol (Goldstein and Brown 1990). The feedback loop includes cell surface receptors, specific for the low density lipoproteins. The stricken individuals lack these macromolecular receptors. As a consequence, the sterol biosynthetic pathway operates at a rate in excess of cellular requirements, and cholesterol accumulates. This, in turn, drives a high incidence of cardiovascular disease, so life expectancy is significantly lowered in individuals homozygous for this condition (i.e., lacking the genes that determine the production of proteins essential to the operation of the negative feedback loop that regulates the activity of HMG-CoA reductase).

Although such examples make clear the deleterious effects of mutations that cause a loss of regulatory properties, it is much more difficult to envisage how those regulatory properties were acquired in the first place. Gene duplication is often invoked as a significant force in evolution, and a covalently elongated version of a catalytic protein carries the possibility for the evolution of allosteric sites (Poorman et al. 1984). Gene fusion could provide similar opportunities (Palm et al. 1985). It is less obvious how regulatory subunits might be recruited.

Biochemical adaptation is a term that covers events taking place on a wide variety of timescales. The evolution of a capacity for metabolic control will take place over many generations. Appropriate metabolic control may, however, make it possible for an organism to adapt to diurnally or seasonally changing conditions within its own lifetime. The flexibility conferred by metabolic regulation enables an organism to move from one environment to another, either as part of its development (metamorphosis) or as part of its feeding activity. On the shortest of timescales, predator-prey relationships are likely to demand adjustments of the rate of energy metabolism in a matter of seconds. On all timescales, biochemical adaptation extends the range of environments that can be survived, and thus widens the life boundary.

It is important to recognize that the ability of an organism to adapt in the physiological sense, to respond appropriately to environmental change on a variety of timescales within its lifetime, is a phenotypic characteristic of that organism. That phenotype stems from the genes that determine the molecular machinery subserving the regulation of metabolism. Such metabolic control is essential for a response which is adaptive, enhancing the survival and reproduction of the individuals possessing that machinery. This enhancement will, in turn, ensure the persistence of the relevant genes in the population.

Gaia, Adaptation, and Selection

This cursory survey of the molecular means that underlie cellular homeostasis, along with the Darwinian perspective on their origin, leads directly to the problems that are raised by the idea of a quasi-organismic control system on a planetary scale. If the biogeochemical pathways of the environment are to be integrated into a global metabolism by analogous feedback loops, there will have to be biomolecular sensors and effectors appropriate to bring about that integration. These sensors and effectors will be the products of genes.

Cellular metabolism demands complementariness amongst the genes of the cell (Dawkins 1982). Organismic metabolism demands complementariness between genes that are expressed in various tissues of the organism. Jackson (1987) has pointed out that even within organisms there can be physiological conflict between differing control mechanisms, especially in unfavorable environments. Nonetheless, it is obvious that such conflicts must find a resolution that permits survival.

It is easy to envisage how selection pressures will eliminate uncomplementary genes and how complementarity will confer selective advantage. Global metabolism would appear to demand complementariness amongst genes of different species. But in this global case, it is not clear how such complementarity would be selected. It is this difficulty that lies at the heart of the resistance of biologists to adoption of the Gaian metaphors.

The proponents of the Gaia concept have been sensitive to the criticism that the idea carries teleological implications:

> Neither Lynn Margulis nor I have ever proposed a teleological hypothesis. Nowhere in our writings do we express the idea that planetary self-regulation is purposeful, or involves foresight or planning by the biota. It is true that our early statements about Gaia were imprecise and open to misinterpretation, but this does not justify the persistent, almost dogmatic criticism that our hypothesis is teleological. (Lovelock 1990)

It is not clear why the proponents are so defensive on this issue. All biologists have to deal with the apparent purposiveness of living organisms. As Jacques Monod (1971) observed, "Objectivity nevertheless obliges us to recognize the teleonomic character of living organisms, to admit that in their structure and performance they act projectively—realize and pursue a purpose." Indeed, from this standpoint, for "Gaia" to qualify as a living organism, it would have to exhibit purposiveness (Falk 1981).

Overall, geophysiologists must be prepared to ask questions about the functional significance of components of the planetary system, and these questions will be difficult, perhaps impossible, to couch in a vocabulary innocent of all teleological implications. Again, biologists have not always been so squeamish. Discussions of alternative evolutionary "strategies" are not held to imply organismic planning sessions. The problem about "Gaia" does not lie in any apparent purposiveness that would be a characteristic of a planetary organism. The problem arises because Darwinian evolution provides a well-substantiated account of the origin of that purposiveness, and it is not clear how that account could apply to the origin of interspecific complementarity on a global scale.

Purposiveness is a feature of a well-adapted physiology. For example, the endocrine cells of the pancreas serve the purpose of stabilizing the blood glucose level in the face of temporally fluctuating food intake. A stable level of glucose in the circulating plasma serves the purpose of maintaining function of the central nervous system, thus permitting the animal to explore new environments. To quote Claude Bernard once again, "The constancy of the internal environment is the condition for free and independent life." Put in the language that I have employed in the earlier part of this chapter, regulation of blood glucose permits adaptation to a wider range of environments. Such an adaptation pushes back the "life boundary," thus permitting a wider exploration of environmental space.

I have deliberately chosen this well-known example of the physiological regulation of blood glucose because throughout the history of the Gaia controversy it has been clear that the concept of homeostasis, adumbrated by the physiologist Walter Cannon and exemplified by the control of blood glucose levels, is integral to Gaia. Complex physiological controls call upon the coordinated activity of the products of many genes expressed differentially in various tissues of the organism. In the case cited, the α and ß cells of the pancreatic islets, the adrenal cortex and medulla, and the anterior pituitary are all major players in the control loop. Current biological thinking supposes such a coordination to arise in a stepwise evolutionary fashion. *At each mutational change, selective pressures favor those phenotypic variants in which better cooperation exists amongst the genes expressed specifically in the relevant tissues.* Adaptation, defined in terms of survivorship of offspring, is greater in those individuals with the most cost-effective control systems. Gaian control systems, in contrast, would have to arise from better cooperation between individuals of differing species. Many biologists consider that the difficulties of constructing a plausible scenario for the evolution of such interspecies cooperativeness make the adoption of Gaian metaphors unhelpful, at best, and, at worst, dangerously misleading.

In the discussion to this point I have stressed the difficulties of integrating the idea of a global control system, Gaia, with the widely accepted biological notions of adaptation and selection. One way of dealing with the problem is to call into question the general validity of these latter two complementary ideas. Such tactics have indeed been used by Gaia proponents (Lovelock 1990; Lovelock 1991; Margulis and Hinkle 1991).

James Lovelock has tended to focus his attack on the concept of adaptation—a notion which he refers to as "anaesthetic" and "dubious." Before and beyond the Gaia debate, others have also noted problems that relate to the definition of adaptedness. Lewontin (1978) and Levins and Lewontin (1985) have stressed the difficulties of providing any measure of better or worse adaptation other than differential survival of descendants ("fitness"). Thus one is led to the allegation that "survival of the fittest" means merely "survival of the survivors." In a contribution to the debate that took place at the American Geophysical Union's Chapman Conference held in San Diego in March, 1988, (referred to elsewhere in this chapter as "the San Diego conference"), Kirchner (1991) underlines the point: "To prove (without tautology) that natural selection works, an independent definition of fitness is necessary." He then attempts to illustrate how such independent definition is possible by reference to the well-known textbook example of the increase of the black melanic form of the pepper moth during the nineteenth-century industrialization of Britain. This is perhaps an unfortunate example. To speak of that particular increase in the frequency of the black phenotype (better camouflaged under the new conditions) as an adaptation of the moth to an environmental shift ignores the complexity of the "environment," the definition of which must include predatory birds, trees, lichens growing (or not growing) on those trees, changes in atmospheric chemistry caused by another species (*Homo sapiens*). The pepper moth is thus almost a perfect example of exactly what Lovelock disliked about the concept of adaptation. In those same conference proceedings Lovelock wrote, "Adaptation is a dubious notion, for in the real world the environment, to which organisms are adapting, is determined by their neighbors' activities, rather than by the blind forces of chemistry and physics alone."

Lynn Margulis, on the other hand, tends to focus her attack not so much on the concept of adaptation but on selection: "In accentuating the direct competition between individuals for resources as the primary selection mechanism, Darwin (and especially his followers) created the impression that the environment was simply a static arena for 'nature, red in tooth and claw' " (Margulis and Hinkle 1991). There are two criticisms implicit here. One concerns the question of competitive advantage as a driving force in evolution. I shall return to that issue later in this chapter. The other criticism concerns the idea of the environment as a relatively constant background against which the evolutionary drama is played out, ignoring the reciprocal effects of the evolving biota on the surface and atmospheric chemistry of the planet through geological time. This second critique integrates well with Lovelock's emphasis that the environment to which organisms adapt (or fail to adapt) includes all of their contemporary biota and the direct and indirect influences of the biota on their physico-chemical milieu: "In no way do organisms simply `adapt' to a dead world determined by physics and chemistry alone. They live in a world that is the breath and bones of their ancestors and that they are now sustaining" (Lovelock 1991).

The Daisyworld Answer and Its Problems

Is it possible to model this planetary interaction of biota and environment over geological time? Given the complexity of the processes involved, the answer would appear to be negative. But, in a series of papers beginning in 1983, Lovelock has presented a model of an imaginary planet with an environment defined by a minimal number of variables and with a biota made up of only a few species. The original "Daisyworld" is a two-dimensional planet with an evolving sun—the luminosity of which is increasing in a monotonic fashion very similar to that of our own sun. If the planet were lifeless, its surface temperature (the single parameter defining the environment of Daisyworld) would increase in parallel with the increasing solar luminosity. But if the planet is given a biotic component of even just two species—in this case, daisies, one dark and one light—the differential growth of the two species can modulate the planetary albedo in such a manner that, over a wide range of solar output, the surface temperature of Daisyworld remains relatively constant (figure 5.5).

The surface temperature of any planet is determined by the balance of incoming thermal radiation from the sun and the sum of the outgoing radiation from the planet (given by the Stefan-Boltzmann dependence on temperature) plus the reflected solar radiation (a function of the planetary albedo). In the Daisyworld case, if the planetary albedo is significantly modified by the extent of coverage by dark daisies (absorbing incoming radiation) and light daisies (reflecting incoming radiation), a simple thermostatic system can be set up using six equations (Watson and Lovelock 1983). The model is capable of considerable elaboration (Lovelock 1992), but for the purpose of this chapter I shall not follow the conclusions that have been drawn from multispecies systems with respect to the question of biodiversity. In the context of an attempt to place Gaia within a generally accepted biological consensus, the simplest version will suffice.

There can be little doubt that Daisyworld succeeds in answering some of the earlier criticisms of the Gaia concept. It does show how "foresight and planning need not be involved to explain automatic regulation" (Lovelock 1991). Why then has it not commanded wider acceptance? The answer turns on three major problems that still have to be addressed.

First, it is unclear whether Daisyworlds in general are stable. In the proceedings of the San Diego conference two participants reported that

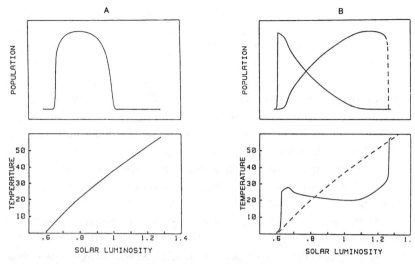

FIGURE 5.5 Lovelock's Daisyworld model. The thermal response of a hypothetical planet, Daisyworld (which is inhabited by just two life forms—a dark and a light daisy) is modeled for a planet which, like Earth, is subjected to an increase in solar luminosity. Case A considers the population changes of the two species taken together as the heat intensifies. Notice that the dark and light daisies do not compete, and the surface temperature of Daisyworld is unaffected by any biotic activity. In Case B, the dark and light daisies not only respond differently to the increase of surface temperature; they alter the albedo of the planet in such a way that global temperature is stabilized over a wide range of solar luminosity.
SOURCE: Figure 1.2 of Lovelock (1991), slightly altered. Reprinted by permission of The MIT Press.

it was relatively easy to construct variants of the Daisyworld model that were unstable. Harte (1991) states, "Of course, it is easy to concoct a model (like Daisyworld) in which the biota dampens disturbance. Unfortunately it is equally easy to concoct models in which the biota enhances disturbance." Chapter 3 showed that current evidence suggests that the chemical activities of the biota associated with the low temperatures of the last ice age may have acted to amplify the temperature shift. If it is true that marine productivity was high in glacial periods, this would imply both an increased sink for CO_2 in surface waters and an enhancement of DMS production. The enhanced withdrawal of CO_2 would draw down atmospheric CO_2 and thus change the radiative balance of the planet in the direction of a more pronounced decrease in temperature. An increase in DMS production would bring a corresponding increase in non–sea salt SO_4^{2-} in the atmosphere, a higher density of particles around which water vapor will condense, and, as a consequence, higher global cloud albedo and decreased global temperature. Both the carbon-sink and albedo effects of hypothesized biological activity during the ice age cooling are, looked at from the standpoint of the physiology of marine algae, quite independent. Looked at from a geophysiological point of view, however, they are synergistic, and it has been conceded that these positive feedback effects are difficult to reconcile with the concept of global environmental homeostasis (Kump and Lovelock 1995). The suggestion by Jenkinson, Adams, and Wild (1991) that the temperature rise predicted as a consequence of the anthropogenic emission of greenhouse gases will be exacerbated by accelerated conversion of soil organic carbon to CO_2 similarly points to a destabilizing effect of the biota on their planetary environment.

The second major problem with the Daisyworld model concerns "optimization." One of the six Daisyworld equations relates the growth rate of the daisies to temperature in a relationship that resembles, graphically, the curve of figure 5.1. In discussing that figure, I pointed out that such bell-shaped curves are a common response of biological systems to environmental variability. However, the value of the environmental variable at which the maximum response is obtained will be subject to natural selection. In organisms (or even populations of the same species) adapted to extreme environments, the maximum will be shifted to a value more appropriate to such environments (Hochachka and Somero 1984: chapter 11). Adaptation on Daisyworld does not, however, occur. The issue of the origin of the two kinds of daisy is ignored; the seeds conveniently (and inexplicably) come to exist on the planet before the temperature has risen to the point that they can sprout. Once they are able to sprout, an equitable temperature is maintained simply by a shift in the populations of the two preexisting color variants.

Overall, the demonstration of environmental homeostasis owing to biotic regulation on Daisyworld is impressive. But it must be noted that such stability is achieved by fixing the value of the environmental optimum ahead of running the model. Evolution on Earth did not have that luxury. Lovelock's claim that "the Daisyworld thermostat has no set point" ignores the fact that an optimum for growth vs. temperature is built into the model. In all the earlier papers on Gaia the term "optimum" or derivatives thereof were frequently used. One of the categories of Gaia hypotheses in the critical taxonomy of Kirchner (1989) is "Optimizing Gaia." Although the claim for optimality appears less frequently in the recent publications on Gaia, Lovelock is quite at ease writing of the Daisyworld system seeking "the most comfortable state" (Lovelock 1991). The problem is not with the anthropomorphism—the problem is that for Lovelock it is the system that makes itself comfortable, "like a cat as it turns and moves before settling" (an ailuromorphism, I suppose). Saunders (1994) puts the matter less colloquially: "regulation appears as a property not of the daisies but of the planet." The contrary view is that, in the course of evolution, genotypes emerge that generate the phenotypes that are "comfortable" in the system. Saunders (1994), arguing in support of Daisyworld, is surely correct when he writes: "It is not in the least difficult to understand why plants should grow best at some particular temperature." However, understanding why real plants (as opposed to imaginary daisies) exhibit optima in their growth-to-temperature relationships is easy only because we know about the opposing effects of temperature on the kinetic characteristics of enzymes and on the stability of these catalysts. Both are functional properties of macromolecular structures read off the plants' genomes and selected to match, roughly, the temperature of their environments. By contrast, the plants of Daisyworld are given their temperature optimum by the modeler.

One of the unstable Daisyworlds presented by a critic at the San Diego conference makes the same point in a different way. Ralph Keeling (1991) introduced a third species of daisy, this one with a higher optimum temperature for growth than the single value attributed to both black and white daisies. In temporal order these new daisies were introduced between the dark and light daisies. This scenario leads to catastrophic extermination and a sterile planet. Of course, to constrain the introduction of daisy species in this way is a deliberate manipulation on the part of the modeler to make a point. Keeling admits this objection but notes that "the original Daisyworld model can be criticized on the same grounds." This is because,

> The regulation of planetary temperature on Daisyworld also requires a specific evolutionary sequence, namely, once the two species of daisy have formed, no further evolutionary changes are allowed. In this sense both the original and modified Daisyworld models are teleological in that they require manipulation of the evolutionary process to produce a desired effect.

Keeling here, correctly, uses "teleological" to refer to the purposes of the model maker as distinct from Monod's use of "teleonomy" to describe the apparent purposiveness brought about by variation and natural selection.

This brings me to the third general objection that can be made to the family of Daisyworld models. Consider, first, that it is unclear that all Daisyworlds are in fact stable. The ones that are stable may be so only because the modeler has introduced an arbitrary characteristic of the daisies (e.g., the modeler has defined the optimum temperature for growth). This latter problem relates to the third and most fundamental objection, namely, that the question of the genesis of the cooperating system is not acknowledged. This question is so important, and so fundamental to what drives criticism of Gaia by biologists, that I shall devote the rest of this chapter to its explication.

How Can a Daisyworld Originate?

The Darwinian answer to the origin of complex metabolic systems lies in the cumulative selection of small stepwise improvements in the chance of survival and reproduction. The argument is compellingly made by writers such as Richard Dawkins (in *The Blind Watchmaker*, 1986) and Stephen Jay Gould (in chapter 12 of *Ever Since Darwin*, 1977). Some such broad generalization would command general assent among contemporary biologists. There would, of course, be vigorous discussion about the tempo of the process and about the importance of disruptive events such as those described in chapter 1. But no matter how smooth or bumpy the progress of evolution might be, and no matter how unpredictable the ensemble of species at any given moment in planetary history might be, two features are necessary components of any evolutionary account. These are (1) sufficient time (measured in terms of biological generation) and (2) a sufficient reservoir of genetic variability.

The chief theoretical weakness of the Gaia concept lies in the absence of any such pool of potential systems from among which "Gaia" could have evolved. It is this weakness that Dawkins recognized in his critique of Gaia: "The universe would have to be full of dead planets whose homeostatic regulation systems had failed, with, dotted around, a handful of successful, well-regulated planets of which Earth is one" (Dawkins 1982:236). He concludes that Lovelock "would surely consider the idea of interplanetary selection as ludicrous as I do."

Daisyworld was indeed conceived by Lovelock as an answer to Dawkins's criticism (Lovelock 1990). But, in my view (and apparently that of many biologists), it misses the point. It is not merely that no evolution occurs *on* Daisyworld, that the genotypes are built into the model at time zero, and that, as pointed out by Keeling (1991), if an evolutionary sequence is added it may well be destabilizing. It is, rather, that Gaia's proponents have offered no causal explanation for the evolution *of* Daisyworld, or, more generally, of any Gaian control system.

The problem is not answered by suggesting that the criticism arises from an ideological "preoccupation with the romantic, Victorian conception of evolution as a prolonged and bloody battle," as claimed by Margulis and Hinkle (1991). That preoccupation may or may not be a problem for evolutionary biology generally. Lynn Margulis herself has done a great deal to foster a fuller appreciation of the role of symbiotic cooperation in evolution. But whatever the relative roles of competitive or cooperative processes in evolution, one cannot evade the underlying requirement that it occurs by selective processes—selecting for cooperative as well as competitive advantage. And none of the proponents of Gaia have adequately addressed this issue of selection.

To repeat, evolution requires a vast reservoir of genetic potential variation on which cumulative selection can operate. It is crucial that the variation be genetic—that is, self-replicable. This aspect of the problem for Gaia has indeed been stressed by Ehrlich (1991). He makes the crucial point that the nonliving, inorganic components of an ecosystem have no means of replication. He illustrates this with the example of transient lakes, such as those of the Australian interior:

> [Lake Eyre] has not found any way to replicate itself, and there is no variant offspring of Lake Eyre that is more resistant to evaporation and able to persist through drought . . . Because they are self-replicating with variation and subject to differential reproduction, organisms can evolve forms that can survive those hideous droughts in the middle of Australia. Lakes, lacking these attributes, have not evolved persistent forms.

On the plus side, it must be noted that Gaia does have one of the prerequisites of an evolutionary scenario—adequate time. There have been living systems on this planet for nearly four billion years. What the Gaia concept lacks is a source of genetic potential variation. A planetary organism (with a continuity of existence at least 80% that of its geophysical substrate) had only one chance to do things right. The possibility of cumulative selection (central to the Darwinian account of the evolution of teleonomic systems) is unavailable at the global scale. Biogenesis had to get it right the first time. Thus, unless one were to turn to arguments from design, the establishment and persistence of life on Earth as a mega-organism becomes an event of vanishingly low probability. Imagine, if you will, how any Daisyworld program could write itself without a James Lovelock.

Attempts have been made to answer even this question. Perhaps, it is suggested, the emergence of Gaia resembles the development of a many-tissued embryo rather than the evolution of a polyphyletic biota. Waddington (1957) referred to the notion of keeping the developmental process relatively invariant as "canalization" or "homeorhesis," as distinct from "homeostasis" in which some physiological state is held constant. Margulis (1990) took up this idea and proposed that Gaia is "more homeorhetic than homeostatic." Others have invoked non-equilibrium thermodynamics to account for the self-organization of a Gaian structure. For a general review of this area see Barlow and Volk (1992), who elsewhere (Barlow and Volk 1990) categorize such suggestions as arguments based on "internal selection."

There are two difficulties attendant to any framework that posits a developmental or internally selected basis for the genesis of Gaian stability and persistence. First, if the origin of Gaia is supposed to be analogous to embryological morphogenesis, it is important to remember how closely that process is specified and restrained *genetically*. Indeed, Waddington introduced the idea (and terminology) of homeorhesis in a 1957 book entitled *The Strategy of the Genes*. Recent advances in molecular genetics are starting to provide the same sort of mechanistic insights into developmental homeorhesis as they have already produced for physiological homeostasis. So such an analogy to embryology reopens the question of how Gaia could evade the requirement for genetic information.

A second problem with a developmental framing of Gaia pertains to the suggestion that Gaia is an example of an open, self-organized system obeying the rules of non-equilibrium thermodynamics. This recasting of Gaia in thermodynamic rather than chemical and biological terms is unhelpful because it proceeds at too high a level of generalization. Despite some theoretical difficulties concerning the decrease of entropy accompanying biogenesis, it is surely true that thermodynamics must permit the emergence of life (thermodynamicists are alive!). Or, as it is put by Kauffman (1995), "Thermodynamics be damned. Genesis, thank whatever lord may be, has occurred. We all thrive." It may be true (though it remains to be seen whether it is more than a truism) that Kauffman is correct when he goes on to show that such emergence, somewhere in the universe, is an inevitable consequence of the laws of complexity. However, even if such laws turn out to be a necessary part of an account of the emergence of any organism, they cannot be sufficient to account for the genesis of a mega-organism, Gaia, on this planet, Earth, at this juncture in the history of the universe. Such generalizations cannot account for the origin of specific assemblages of cells and tissues such as those that constitute the organisms of the biosphere at any given point in Earth's history. It may be that deep sources of order underlie and constrain the particularities and quiddities of biological diversity. Nonetheless, it is the historical contingencies of Darwinian selection operating upon genetic variation that provide credible scenarios for "the origin of species."

Kauffman (1995) and others before him refer to such credible scenarios as evolutionary "just-so stories." Perhaps such disparagement is fair comment. But because we do not have even a "just-so story" for the origin of Gaia (or for the origin of Daisyworlds), it would appear, to return to Weyl's list of bases for stability (discussed in chapter 3), that geophysiological feedback by the biota as a means of restraint within the life boundary becomes merely another sort of "luck." Ehrlich, in his 1991 comments on Gaia, put the matter bluntly, "In fact, if I were to propose a non-Gaia hypothesis, it would be that life has been extremely fortunate. Serendipitously, a lot of negative feedback loops have developed that have so far prevented *Homo sapiens* or life itself from disappearing."

Overall, in the absence of any evolutionary explanation of how Earth could acquire the homeostatic characteristics of an organism, the proponents of Gaian hypotheses are also thrown back on "luck"—however much they may resist the characterization. Ehrlich goes on, "The challenge for those who wish to support the strong version of the Gaia hypothesis is to discover some process by which the planet could have accumulated homeostatic mechanisms that favor life."

6
Molecular Regulation and Global Metabolism

The argument of the previous five chapters has brought us to a quandary. The problem arises because of the contradiction between, on the one hand, the clear indications of a role for the biota in engendering the habitability of the global environment and, on the other hand, the absence of any identifiable process that would explain how the biota could acquire the potential for such a role.

Can Global Stability Be Linked to Physiological Homeostasis?

It is important to recall that the suggestion that the biota have modified the character of our planet in ways that have increased its amenability to life is only a partial answer to the question of habitability. Of course, as we saw in chapter 3, the habitability of Earth is made possible by cosmic considerations—its size, its chemistry, its distance from a parent star with the "right" luminosity. Many such factors provide necessary conditions for the genesis of living systems possessing a polynucleotide-based genetic code and a polypeptide-based expression of the genetic message. The structure and function of DNA, RNA, and the proteins demand relatively stringent physico-chemical conditions; it is only on a planet where such conditions obtain that life as we experience it can get started, persist, and flourish.

It appears that neither self-organization nor selection have been able to get started on any of the other bodies of the solar system. Thus, the

Molecular Regulation and Global Metabolism 147

problem of habitability will remain, whatever the final assignment of evolutionary roles to the propensity of complex systems to self-organize or to the processes of selection and adaptation. But, as we also saw in chapter 3, there is ample evidence that the planetary biota do play major roles in the feedback loops that stabilize the characteristics of Earth's atmosphere and hydrosphere.

It thus appears that, because stability is a prerequisite for habitability, the biota must be included in any catalog of the conditions that have permitted the maintenance of habitability on Earth. Furthermore, the particular ensemble of individuals and species that exists on the planet at any time in its history influences not only the stability of the characteristics of the planetary environment but also the contemporaneous levels, the numerical values, of those characteristics. The steady-state level of any such environmental parameter is determined by many factors, but the list of such causal influences must include, alongside the physics and chemistry of Earth's crust, the biochemistry and biophysics of those organisms that participate in the processes that determine the value at which that parameter is set.

It is not a difficult leap from such consideration of the role of the biota in setting and stabilizing the characteristics of the global environment to the attractive metaphor of the planetary organism "Gaia." In the Gaian simile the feedback loops that determine, for example, the oxygen content of the atmosphere are seen as analogous to the physiological feedback loops that set and stabilize such characteristics of the internal environment of metazoan animals as the glucose content of the blood. But it is in drawing out the implications of such a "geophysiology" (Lovelock 1986) that one encounters a second facet of the problem. The cumulative occurrence of mutation and selection provide an evolutionary process by which advantageous physiological homeostasis could arise. By contrast, as we have seen in chapter 5, there is no known "process by which the planet could have accumulated homeostatic mechanisms that favor life" (Ehrlich 1991).

A possible solution to the fundamental obstacle calling Gaian metaphors into question would arise if internal physiological control mechanisms, the genesis of which is accounted for by accepted Darwinian theory, also served to set external environmental parameters within the life boundary. Such an answer to our problem could arise in one of two ways. One answer turns on chance. The other looks to inevitability.

First, it might be that some particular mechanisms of physiological homeostasis, selected to meet teleonomic needs of organisms, may also, by chance, determine and stabilize key external environmental parameters. This would be merely a fortuitous by-product and thus another instance of the role of "luck" in permitting the establishment and persistence of living systems on Earth. Such chance occurrence cannot be ruled out, and so throughout the rest of this book it may be well to bear in mind that, even if an inevitable causal linkage between internal and external feedback loops cannot be established, there may be cases where the involvement of the biota in global environmental regulation does occur via the sensors and effectors of internal regulatory mechanisms. It will be important to identify such cases.

The second way in which the problem might be resolved stems from an understanding of the way in which the biota determine and stabilize environmental characteristics. The values of reservoir size and turnover time define a steady-state achieved by the balance of the global processes referred to in the four-box scheme of figure 1.6 as "assimilation," "regeneration," "mobilization," and "sequestration." At the organismic level these processes are closely coordinated. Variants in which the flow of material and energy is better coordinated will, *ceteris paribus*, possess greater Darwinian fitness than those in which flows are not so well adjusted. Such coordination is a phenotypic character read off from complementary genes in a genome that has been subject to natural selection. Is it possible that such intraorganismic coordination of biochemical processes is responsible for their global balance, which will in turn set values of the corresponding geochemical parameters? And, crucially, is this global balance achieved not merely by happenstance but as an inevitable consequence of the involvement of biomolecular regulatory processes in the global biogeochemical cycles?

A positive answer to these questions would provide a resolution to the central problem that this chapter addresses—the absence of underlying mechanisms that could subserve the roles that the biota play in keeping global environmental variables within the life boundary. It may be that the problem could be resolved if one could present cases where cellular regulation of metabolism, whether "crystallized" (Kauffman 1993) or selected during the evolutionary process, can be demonstrated to play a part in determining the flux rates in the global biogeochemical cycles. For instance, is it possible to bridge the gap between what is known of the coordination of the elemental cycles at the ecological level and our knowledge of the molecular control of the flow of these same elements through metabolic pathways?

Liebig's Law and Redfield's Ratios: Coordination at the Ecological Level

In the four-box scheme presented in figure 1.6, the chemical element under consideration is not identified. This is because the four-box scheme is intended to be generic—applicable to any of the elements that cycle through (and are therefore crucial to) life. Four elements that are universal and major cellular constituents and that are commonly thought of as "nutrient" are carbon, nitrogen, phosphorus, and sulfur. The assimilation pathway of the four-box scheme of, say, carbon must be accompanied by the incorporation of nitrogen, phosphorus, and sulfur. Correspondingly, in any such representations of the global cycles of nitrogen, phosphorus, and sulfur, the rates of assimilation in these cycles must be correlated to one another and to that of carbon.

At the ecological level two important and general correlative rules have been widely accepted. First, primary productivity will be limited by the availability of one of these four key elements (or by the availability of some other element essential for autotrophic growth). Second, these four elements will be incorporated into biomass in fixed proportions.

The dependence of growth on a limiting nutrient was clearly formulated by Justus Liebig in 1855: "The growth and the development of a plant depend on the assimilation of certain bodies, which act by virtue of their mass or substances. This action is within certain limits directly proportional to the mass or quantity of these substances" (Liebig 1855). One might question Liebig's assumption of direct proportionality. The relationship between the rate of a biological process and the availability of the substrate for that process is not, in general, linear. The relationship for dependence of growth on nutrient concentration is thus more likely to resemble the well known Michaelis-Menten curve (figure 6.1) in which the order of an enzymic reaction with respect to substrate varies from 1 at low substrate concentration (i.e., the rate is directly proportional to substrate concentration) to 0 at high substrate concentration (i.e., the rate is independent of substrate concentration). The algebraic expression of the relationship requires two constants: V_{max}, which is the velocity of the enzyme-catalyzed reaction extrapolated to infinite substrate concentration, and the Michaelis constant (K_M), which measures the "affinity" of the enzyme for its substrate.

Microbial growth is known to obey such a concentration-dependent relationship to limiting nutrient (Monod 1942), and similar nonlinear

150 *Molecular Regulation and Global Metabolism*

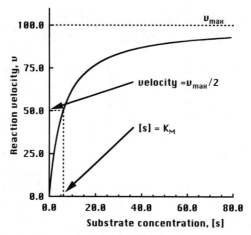

FIGURE 6.1 The Michaelis-Menten relationship between the rate of an enzymic reaction and the substrate concentration. A similar relationship holds between microbial growth and nutrient concentration. V_{max} is the velocity of the enzyme-catalyzed reaction extrapolated to infinite substrate concentration. K_M is equal to the substrate concentration at which the reaction velocity = $V_{max}/2$. It is a measure of the affinity of the enzyme for its substrate.

relationships may hold for a wide range of biological systems. For this reason, many models of biogeochemical cycles assume a fractional order dependence of growth on nutrient concentration. One such example is the "biota growth factor," used by Bacastow and Keeling (1973), of a magnitude greater than zero but less than one. If one models a biogeochemical cycle in which such a process is involved, then both the steady-state characteristics and the dynamic behavior of the cycle will be affected by the values of V_{max} and K_M. The nonlinear nature of this relationship generates a certain amount of mathematical complexity, but the biota can still be thought of as acting as simple catalysts for the process. Different systems will vary only with respect to the values of V_{max} and K_M associated with the biocatalytic process. The values of these constants may, however, be significant for global metabolism, as we saw in the discussion (chapter 3) of how the level of atmospheric O_2 is determined.

There is nevertheless an important difference between growth and a

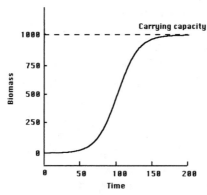

FIGURE 6.2 The Verhulst relationship between growth and time. Exponential growth is limited at some environmentally imposed carrying capacity.

simple enzymic reaction. Growth is autocatalytic; that is to say, the catalyst for growth is also the product of growth. Such an autocatalytic process is explosive and would, if unrestrained, lead to an infinite biomass. One way in which the assimilation pathway of the four-box scheme is restrained is by exhaustion of one of the four boxes: the nutrient pool. Any attempt to model the dynamic behavior of biogeochemical cycles must take this problem into account. One way of doing so is to provide an upper limit for growth, as in the density-dependent equation for population growth depicted in figure 6.2. The upper limit is the carrying capacity. This logistic equation, named after Pierre-François Verhulst, contains both a rate-constant for growth and a limiting upper level for population. For biogeochemical purposes, this upper limit would be fixed in terms of biomass rather than population. This upper limit, or carrying capacity, is however an ad hoc factor inserted to make the model behave properly and to evade the problem of autocatalytic growth. A general model of biogeochemical cycling cannot merely assume restraint of growth; one of the criteria for successful modeling is the ability to demonstrate such restraint.

One possible factor that could limit the rate of incorporation of one element into biomass is the availability of another element. Such limitation is the basis of the ecological generalization that has come to be known as Liebig's Law of the Minimum. This law was formulated by Liebig in 1855. It is the combination of two of his fifty propositions:

Proposition 40. The absence or deficiency, or the want of available form in that one constituent, renders the others which are present ineffectual, or diminishes their efficacy.

Proposition 41. By the deficiency or absence of one necessary constituent, all the others being present, the soil is rendered barren for all those crops to the life of which that one constituent is indispensable.

These two propositions have been widely accepted within ecology. It is noteworthy that Proposition 40 is less restrictive than Proposition 41. The latter suggests a complete cessation of growth as the result of a deficiency in one element, whereas Proposition 40 speaks of the deficiency of one essential constituent diminishing the efficacy of the others.

It is relatively simple to combine one four-box representation of the cycle of one element with that of another element and to show mathematically the dependence of the assimilation pathway on the two nutrient pools. Appropriate adjustments of the Michaelis constants, K_M, for the dependence of assimilation on each of the nutrients can give rise to cases that seem to exemplify Liebig's Law. In these cases growth may be a function of just one nutrient and independent of the other. It is, however, possible to adjust the values of these constants to demonstrate that such manifestations of independence are limiting cases and that a full range of intermediate cases is computable in which the concentrations of both of the nutrient elements will appear in the rate equation for assimilation (Williams 1987). The significance for real world ecology of such simplistic mathematical arguments is not clear. It does however suggest, for instance, that the long-standing debate among oceanographers as to whether the growth of marine phytoplankton is limited by nitrogen or phosphorus could be resolved by the recognition that both elements may play a part simultaneously in determining this growth.

A second rule concerning the correlation of rates of assimilation in the cycles of different elements is an acknowledgment of the relative constancy of the proportions of C, N, P, and S in organisms. Alfred Redfield (1934) proposed a constant stoichiometric relationship between C, N, and P in marine phytoplankton. It is now recognized that these "Redfield ratios" may not be applicable to terrestrial systems (Likens et al. 1981; Hunt et al. 1983), and there has been some debate as to how constant this ratio is in marine plankton (Peng and Broecker 1984). The cause of these discrepancies is, however, fairly obvious and need not diminish the validity of Redfield's proposal. Consider: many terrestrial organisms build up large amounts of structural materials or energy reserves. Typically these will be polysaccharides, lignins, or lipids containing little or no nitrogen or phosphorus. It would be confusing to attempt to eliminate these structures or reserves from the accepted definition of biomass. But the four-box scheme used throughout this book offers a way out. Here, the biomass component of a tree is equated to its living tissue; its woody parts are assigned, rather, to the bioproduct reservoir. A similar argument was used in chapter 1 in seeking to estimate the gross rate at which inorganic nitrogen is incorporated into organic forms.

All cellular material has a relatively uniform composition with respect to its major constituents (proteins, nucleic acids, carbohydrates, and structural lipids). This means that, leaving inert structures and reserve materials aside, there must be severe restraints on the elemental composition of the active cellular material. A constancy of the elemental composition implies that the rates of assimilation of C, N, and P must be in the same proportion to one another as the stoichiometric ratios of the elements represented in the biomass compartment. It is, however, relatively easy to model situations in which this restriction of the relative rates of assimilation of nutrient elements conflicts with Liebig's Laws (Williams 1987). Somehow, these macroscopic, ecological rules of correlation of flow rates from nutrient to biomass pools must be reconciled at the biochemical level.

Liebig's generalizations can be seen to have their underpinnings in the law of mass action. Determination of primary productivity by pCO_2 can be understood in terms of the general relationship between rate of reaction and substrate concentration (the Michaelis-Menten equation). On the other hand, an explanation of the control of primary productivity in the carbon cycle by the availability of NO_3^- or NH_3 is more likely to have to invoke specific biochemical feedback processes that reside within cells. There is now, in fact, considerable physiological evidence for the existence of such intracellular controls. It has been shown that the metabolic flow of carbon in photosynthetic organisms is profoundly affected by concomitant processes of nitrogen assimilation (e.g., Elrifi et al. 1988; Turpin et al. 1990). Such interactions imply the existence of molecular machinery capable of effecting such integration. This is, perhaps, a promising area in which to seek instances of the global significance of molecular regulatory mechanisms. It is probably not possible to demonstrate such significance unequivocally, given our rudimentary knowledge of global biogeochemical cycles. Rather, in this chapter, some of the biochemical regulatory processes of one particular cycle, that of nitrogen, will be reviewed with the more modest aim of establishing the *reasonableness* of postulating geochemical significance for molecular aspects of biological nitrogen metabolism.

The Molecular Basis of the Global Nitrogen Cycle

Figure 6.3 adapts the generic four-box scheme of chapter 1 to call attention to the three key enzymes of the nitrogen cycle that I will discuss in detail later in this chapter. Note that the three enzymes are, of course, found exclusively in living organisms and, therefore, in one sense belong in the *biomass* box. This feature of biogeochemical cycling has been alluded to earlier in this chapter when I discussed autocatalysis and the Verhulst equation. Many of the key processes of the biogeochemical cycles are in fact catalyzed by biomass. Note also that, from an ecological point of view, both nitrate (NO_3^-) and ammonia (NH_3 or NH_4^+) are environmental sources of nitrogen that can be converted into biomass. Therefore, both of these forms of "fixed nitrogen" must be placed in the *nutrient* box. Under aerobic conditions ammonia will be oxidized to nitrate, which is the dominant form of inorganic nitrogen in aerobic soils and the preferred nitrogenous nutrient for many plants. But there is no enzyme that converts nitrate directly to organic forms of nitrogen. Nitrate must first be reduced to ammonia before incorporation into the nitrogen-containing molecules of biomass. This reduction, catalyzed by nitrate reductase, as well as the oxidation of ammonia to nitrate are included in the nutrient box.

As discussed in chapter 1, the processes of nutrient interconversion, both of denitrification and nitrification, produce atmospheric by-products: nitrous oxide and dinitrogen. These side reactions are labeled in figure 6.3 as *segregation*, an exit from the active biological cycle. For most organisms, the abundant gas dinitrogen is inert. A small subset of the global biota possess the enzyme nitrogenase, and are therefore enabled to convert dinitrogen to ammonia. This process is labeled as *mobilization*. Most of the global regeneration of nutrient nitrogen must take place via the decomposition of plant biomass. The excretion of animal products and the regeneration of nutrient by decomposition of these products is not a negligible step in biogeochemical cycling—as is witnessed by the importance of animal manure in traditional agriculture practice and the high content of nitrogen in seabird guano deposited on Pacific islands.

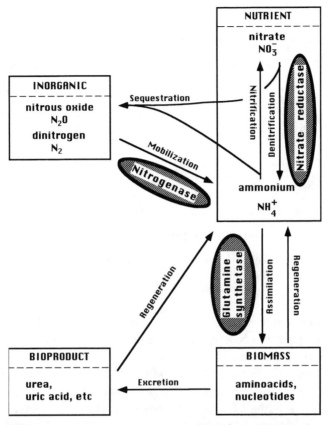

FIGURE 6.3 Key reservoir constituents and enzymic reaction pathways of the nitrogen cycle, as displayed in the four-box format that was generically presented in figure 1.6.

Quantitative estimates of the global nitrogen flow through the pathways of figure 6.3 have been given in chapter 1. The dynamics of the small pool of atmospheric nitrous oxide (N_2O) are crucial to a quantitative understanding of the global nitrogen cycle. The atmosphere contains 300 ppbv of N_2O, equivalent to 53.6 Tmoles N_2O, 107 Tatoms N (Bolle et al. 1986). The only known process for the removal of N_2O is photolytic destruction in the atmosphere. This oxidation to nitric oxide (NO) takes place at a rate of about 0.65 Tatoms/y; the half-life of the pool of atmospheric N_2O is therefore about 114 years. If the N_2O of the atmosphere is in a steady-state, then its destruction rate must be balanced by a production rate of 0.65 Tatoms/y. The major source of N_2O is biological; it is a product of both denitrification and

156 *Molecular Regulation and Global Metabolism*

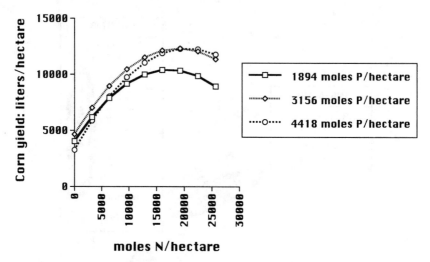

FIGURE 6.4 Corn yield affected by the application of fertilizer nitrogen at various levels of phosphorus.
SOURCE: The data are taken from Munson and Doll 1958.

nitrification (figure 1.12). The other major gaseous product of biological denitrification is N_2, the total flux of which cannot be accounted for because the reservoir of N_2 in the atmosphere is so large that its biological inputs do not measurably affect it. Thus, the lowest possible limit for the loss of nitrogen to the atmosphere from ammonia (NH_3) and from nitrate (NO_3^-) is 0.65 Tatoms/y. The flux value noted in chapter 1 is 10 Tatoms/y for terrestrial systems alone. Rosswall (1983) gives the size of the soil inorganic pool as 11 Patoms and that of the oceanic pool as 41 Patoms. Using a minimal rate of loss of fixed inorganic nitrogen equal to the production rate of N_2O, an upper limit for the half-life of the pool of fixed nitrogen ($NH_3 + NO_3^-$) is calculated as 55 Ky.

On a geological timescale 55,000 years is a very brief period. Such a short half-life for loss of the nutrient pool would suggest an instability which must be counteracted by an equally active process of N_2 fixation. A pool of available inorganic N is essential to the maintenance of the global biota. The extraordinary gains in agricultural productivity brought about by the application of nitrogenous fertilizers (figure 6.4) suggest that much of global biospheric growth is limited by the availability of fixed nitrogen. Any prolonged failure to balance the losses caused by denitrification would be catastrophic for the global biota. In the absence of perturbation (either recently by human activity, as discussed in chapter 2, or on a geological timescale, as discussed in chapter 1) the biochemical processes of figure 6.3 must be balanced. Do the molecular systems that subserve these processes have properties that are important in achieving environmental homeostasis?

The key enzymes emphasized in figure 6.3 are glutamine synthetase (assimilation), nitrogenase (mobilization), and nitrate reductase. These three are not, of course, the only enzymes implicated in the biologically mediated nitrogen cycle. Other enzymes also play important roles, and there are other macromolecular components involved in the complex intra- and inter-cellular interactions that underlie, for instance, the host-symbiont relationships that are so important for nitrogen fixation. This chapter will be confined largely to these three enzymes, partly because of their key roles and partly because their regulation is becoming well understood. In any case, they suffice to make my point about the way in which the analysis of regulation at the molecular level may disclose the kind of integrative mechanisms that would substantiate the metaphor of global metabolism.

The Molecular Control of Glutamine Synthetase

Quantitatively, the most important of the three enzymic reactions is that catalyzed by glutamine synthetase. By way of glutamine synthetase, the nitrogen of ammonium is incorporated into the amide group of glutamine, at the expense of the free energy of hydrolysis of the terminal pyrophosphate bond of ATP:

$$\begin{array}{c}
\text{COO}^- \\
| \\
{}^+\text{H}_3\text{N–CH} \\
| \\
\text{CH}_2 + \text{NH}_4^+ + \text{ATP} \\
| \\
\text{CH}_2 \\
| \\
\text{COO}^-
\end{array}
\xrightarrow{\text{glutamine synthetase}}
\begin{array}{c}
\text{COO}^- \\
| \\
{}^+\text{H}_3\text{N–CH} \\
| \\
\text{CH}_2 + \text{ADP} + \text{P}_{\text{inorg}} \\
| \\
\text{CH}_2 \\
| \\
\text{C=O} \\
| \\
\text{NH}_2
\end{array}$$

glutamate glutamine

158 *Molecular Regulation and Global Metabolism*

The 5-amide of glutamine then becomes the source of nitrogen for all the other complex organic nitrogen compounds of cells and tissues. Thus, virtually all of the 1.17 Patoms of nitrogen that are incorporated annually in the formation of biomass are channeled through glutamine, the synthesis of which is catalyzed by glutamine synthetase. This generalization includes the 2-amino group of the acceptor, glutamate. Glutamate is itself synthesized in the glutamate synthase reaction:

$$\begin{array}{c}\text{COO}^-\\|\\ \text{C}=\text{O}\\|\\ \text{CH}_2\\|\\ \text{CH}_2\\|\\ \text{COO}^-\end{array} + \begin{array}{c}\text{COO}^-\\|\\ {}^+\text{H}_3\text{N}-\text{CH}\\|\\ \text{CH}_2 + \text{NADPH}\\|\\ \text{CH}_2\\|\\ \text{C}=\text{O}\\|\\ \text{NH}_2\end{array} \xrightarrow{\text{glutamate synthase}} 2 \begin{array}{c}\text{COO}^-\\|\\ {}^+\text{H}_3\text{N}-\text{CH}\\|\\ \text{CH}_2 + \text{NADP}\\|\\ \text{CH}_2\\|\\ \text{COO}^-\end{array}$$

2-oxoglutarate   glutamine   2 glutamate

The reductant in this reaction may be either reduced nicotinamide adenine dinucleotide phosphate (NADPH) or reduced ferredoxin, depending upon the source of the enzyme. The sum of the glutamine synthetase and glutamate synthase reactions may be written:

$$\text{2-oxoglutarate} + 2\text{NH}_3 + 2\text{ATP} + \text{reductant} \rightarrow \text{glutamine} + 2\text{ADP} + 2\text{P}_{\text{inorg}} + \text{oxidant}$$

Oxoglutarate is a key intermediate in the cellular metabolism of lipids and carbohydrates. Glutamine is the entry point into the formation of the many cellular constituents that contain nitrogen. This reaction sequence serves as the molecular link between the processes that incorporate the inorganic forms of the nutrients carbon and nitrogen into biomass (figure 6.5). Thus, "the linkage of carbon and nitrogen that is seen in global biogeochemical cycles has a basis at the level of cellular biochemistry" (Schlesinger 1991:143). More specifically one might assert that it is the molecular machinery of the glutamine synthetase reaction, operating on the atomic scale, that ties together the planetary-scale

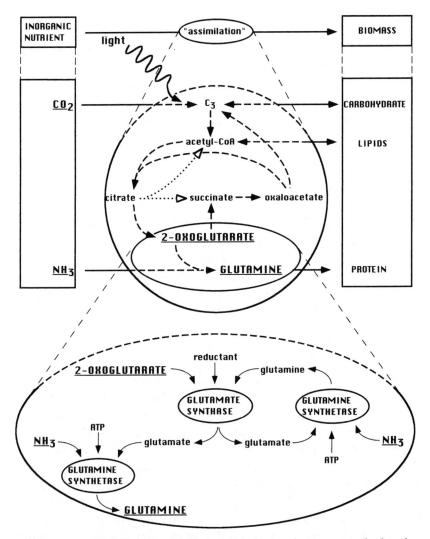

FIGURE 6.5 The ecological assimilation of carbon and nitrogen probed to the level of metabolic pathways and to the enzymic level. Figures 6.7 and 6.9 will probe the control of carbon and nitrogen assimilation more deeply, giving close-ups of the activation of glutamine synthetase (figure 6.7) and of the operon responsible for the synthesis of glutamine synthetase (figure 6.9). The enzyme common to these two processes is P_{II}. Figure 6.6 carries the probing one step further to show the response of P_{II} as receptor of the metabolic signals that link the cycles of carbon and nitrogen.

tion, operating on the atomic scale, that ties together the planetary-scale biogeochemical cycles of carbon and nitrogen.

A detailed understanding of the control of the glutamine synthetase reaction will be essential in any attempt to describe the molecular regulation of the global nitrogen cycle. Unfortunately, no such comprehensive picture can yet be drawn for the primary producers that drive most of the assimilation of nitrogen into biomass. In recent years, however, a remarkably detailed analysis has been made of the control of glutamate synthetase in the enteric bacteria.

On the assumption that the general principles of these control mechanisms are likely to be broadly applicable, this analysis may serve as a paradigm for the molecular regulation of global metabolism. The regulation of glutamine synthetase has been reviewed by Rhee et al. (1989). The regulation displays all the patterns of control discussed in chapter 4—modification of enzyme activity, both by reversible binding of low molecular weight ligands and by the modification of specific residues in enzymic cascades, as well as by changes in the level of transcription of those segments of the genome coding for the polypeptides involved in the system.

All such regulatory systems must possess some way of sensing the relevant environmental variables. It appears that the primary sensor in the regulation of glutamine synthetase is a uridylyltransferase. This transferase is capable of modifying a regulatory protein, P_{II}, in one of two ways: (1) by the transfer of a uridylyl group from UTP into a phosphodiester linkage with a specific tyrosine residue in each of the four identical subunits of P_{II}, or (2) by the hydrolysis of these uridylyl-tyrosine bonds to yield UMP and unmodified P_{II}. Both the uridylylation and the deuridylylation reactions are catalyzed by the same protein (Garcia and Rhee 1983). Because this is a catalytic reaction, one molecule of the transferase/hydrolase can bring about the modification of many molecules of P_{II}. Significantly, the balance between uridylylation and deuridylylation is allosterically regulated by the levels of 2-oxoglutarate and glutamine—the input and output of the reaction sequence that connects carbon and nitrogen metabolism. The uridylylation is activated by 2-oxoglutarate and is inhibited by glutamine, whereas the hydrolase reaction (deuridylylation) is activated by glutamine (figure 6.6).

This responsiveness of the bifunctional uridylyltransferase to 2-oxoglutarate and glutamine (2-OG and GLN in figures 6.6 to 6.8) triggers a complex chain of events. As figure 6.6 suggests, when the ratio of 2-oxoglutarate and glutamine is high, P_{II} will exist predominantly in the

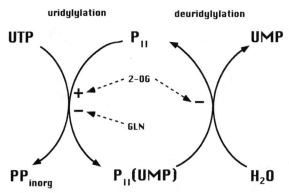

FIGURE 6.6 The uridylylation/deuridylylation cycle of the regulatory enzyme P_{II}. This cycle is the initial process in the response of the glutamine synthetase system. UTP is uridine triphosphate; UMP is uridine monophosphate; 2-OG is 2-oxoglutarate; GLN is glutamine; PP_{inorg} is pyrophosphate.

modified (uridylylated) form (P_{II}-UMP). But when the ratio is low, the deuridylylation reaction will be favored, and P_{II} will exist largely in the unmodified form. In turn, P_{II} is capable of changing both the genetic expression of glutamine synthetase (the synthesis of new enzyme) and the catalytic activity of already existing enzyme. The direction of the changes depends upon the uridylylation state of P_{II}, and in both cases the effects involve intermediate enzymically catalyzed modifications.

Let us turn first to an explanation of the effect of P_{II} on the catalytic activity of glutamine synthetase. Glutamine synthetase exists as a dodecamer of identical subunits arranged as a double-decked hexagon. Each subunit is subject to modification by the transfer of an adenylyl group from ATP to the tyrosine residue at position 397 of its polypeptide chain (Colombo and Villafranca 1986). The fully modified form will contain 12 adenylyl groups, but intermediate degrees of adenylylation are possible. As in the case of the modification of P_{II}, both the adenylylation and deadenylylation of glutamine synthetase are brought about by a single protein (shown as AT in figure 6.7) (Caban and Ginsburg 1976). These two catalytic reactions are influenced by the modified and unmodified forms of P_{II}. The deadenylylation reaction exhibits an absolute requirement for modified P_{II} (P_{II}-UMP). Conversely, the transfer of the adenylyl groups from ATP to give modified glutamine synthetase, GS-$(AMP)_{12}$ in figure 6.7, is stimulated by unmodified P_{II}. The relationships to the signal are as follows:

162 *Molecular Regulation and Global Metabolism*

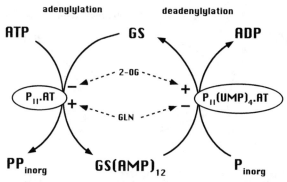

FIGURE 6.7 The adenylylation/deadenylylation cycle of glutamine synthetase. This cycle activates and deactivates glutamine synthetase (GS). Note the two different forms of P_{II} and their differential effects on the pathways of adenylylation and deadenylylation.

2-OG/GLN high $\therefore$ $P_{II} \to P_{II}$-UMP $\therefore$ GS(AMP)$_{12} \to$ GS

2-OG/GLN low $\therefore$ P_{II}-UMP $\to P_{II}$ $\therefore$ GS $\to$ GS(AMP)$_{12}$

Superimposed upon the regulation of the adenylylation and deadenylylation reactions by P_{II} and P_{II}-UMP are allosteric effects of 2-oxoglutarate and glutamine. As shown in figure 6.7, 2-oxoglutarate inhibits the adenylylation reaction and stimulates the P_{II}-UMP–dependent phosphorolysis of the adenylyl-tyrosine-397 bond. Glutamine, reciprocally, inhibits the P_{II}-UMP–dependent deadenylylation and stimulates the transfer of adenylyl groups from ATP to tyrosine-397. These allosteric effects of 2-oxoglutarate and glutamine on ATP:glutamine synthetase adenylyl transferase (AT) augment the effects mediated via the uridylylation and deuridylylation reactions of P_{II}.

The functional significance of these covalent modifications and allosteric effects arises because of the marked differences in kinetic properties of adenylylated and deadenylylated glutamine synthetase. Glutamine synthetase is subject to end-product inhibition (cf. figure 4.1). Because it is the enzyme responsible for the initial step incorporating nitrogen into cellular metabolism, glutamine synthetase stands at the beginning of many biosynthetic pathways. Consequently, it is not surprising that it is subject to inhibition by such end-products as CTP, AMP, histidine, tryptophan, alanine, glycine, serine, and carbamyl phosphate. The effect of adenylylation is to make the enzyme much more sensitive to these allosteric inhibitors and to change its divalent cation requirement from Mg(II) to Mn(II). The result of these changes is that, under normal physiological conditions, the adenylylated enzyme would be inactivated.

Partially adenylylated enzyme would retain activity in proportion to the number of unmodified subunits. Functionally, the control logic follows from the relationships just discussed:

$$2\text{-OG/GLN high} \therefore GS(AMP)_{12} \rightarrow GS \therefore 2\text{-OG} \rightarrow GLN$$

$$2\text{-OG/GLN low} \therefore GS \rightarrow GS(AMP)_{12} \therefore 2\text{-OG} \nrightarrow GLN$$

The predicted relationship between the adenylylation state of glutamine synthetase and the intracellular ratio of 2-oxoglutarate to glutamine was observed in vivo by Senior (1975). Because of the activity of partially adenylylated enzyme, it might be expected that the biological advantage conferred by the high amplification produced by these cascades is not so much the production of an "on-off" switch but, rather, the ensuring of rapid fine tuning of enzyme activity. On the other hand, the findings of Senior (1975), as shown in figure 6.8, suggest that there may be critical values of intracellular metabolites around which small changes in their ratio may produce dramatic changes in adenylylation.

These effects on the *activity* of glutamine synthetase molecules already existing in the cell are complemented by effects on the *synthesis* of glutamine synthetase. This causative chain also begins with the uridylylation and deuridylylation of P_{II}. This protein is capable not only of influencing the activity of ATP:glutamine synthetase adenylyl transferase (as portrayed in figure 6.7). In its deuridylylated form (though not when modified by the addition of uridylyl groups), it activates the phosphatase activity of a protein, NR_{II}, which was discovered as the product of the *gln* L (*ntr* B) gene (Ninfa and Magasanik 1986; Keener and Kustu 1988). This enzyme (NR_{II}) is bifunctional. In the absence of deuridylylated P_{II}, it acts as a protein kinase, transferring phosphate from ATP (probably via a phosphohistidine residue in its own active site: autophosphorylation) to an aspartate residue (Weiss and Magasanik 1988) in NR_I, the product of the *gln* G (*ntr* C) gene (figure 6.9). NR_I is a response regulator that, in its phosphorylated form, NR_I-P, binds to DNA sequences upstream from (and thus activates) the *gln* operon which includes the structural gene for glutamine synthetase (*gln* A). The control logic is therefore:

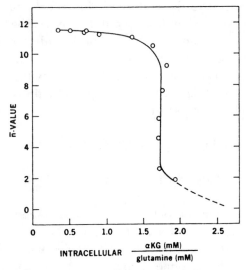

FIGURE 6.8 The in vivo variation of the extent of adenylylation of glutamine synthetase with the ratio of 2-oxoglutarate (αKG in this figure) to glutamine. These findings directly link the molecular to the cellular level. Source: Figure 8 of Senior 1975. Copyright © American Society of Microbiology 1975. Reprinted by permission of the American Society of Microbiology.

2-OG/GLN high ∴ $P_{II} \rightarrow P_{II}$-UMP ∴ $NR_I \rightarrow NR_I$-P ∴ *gln* A active

2-OG/GLN low ∴ P_{II}-UMP $\rightarrow P_{II}$ ∴ NR_I-P $\rightarrow NR_I$ ∴ *gln* A inactive

Thus, the deuridylylation of P_{II} not only sets off a sequence of events that leads to the inactivation of glutamine synthetase; it also, by a separate sequence, inactivates the gene responsible for the production of the enzyme. In both cases an enzymic cascade is involved—the activity of a single molecule of the uridylylating/deuridylylating enzyme will transform many molecules of P_{II}. These in turn will activate either the adenylylating/deadenylylating enzyme (AT) or the kinase/phosphatase (NR_{II}). AT will then modify many molecules of glutamine synthetase; NR_{II} will modify the *gln* operon response regulator. A single active copy of the *gln* structural gene can, in turn, direct the production of many transcripts. Three or four levels of amplification therefore happen between the molecular effect of the 2-oxoglutarate/

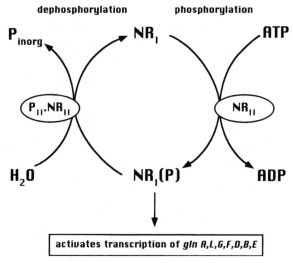

FIGURE 6.9 The phosphorylation/dephosphorylation cycle of the bifunctional protein NR_I and the resultant effect on the *gln* operon.

glutamine ratio on the uridylylation state of P_{II} and the end effect on the rate of glutamine production.

The control of bacterial nitrogen metabolism belongs to a common pattern of regulatory mechanisms (the "two-component pathway") involving polypeptides with conserved sequence similarities (Stock et al. 1990). However, it should be reiterated that the particular ways (just discussed) for controlling nitrogen assimilation are known to be in effect only in certain procaryotes (the enteric bacteria); they may not be characteristic of the major primary producers in the global cycle of nitrogen. Indeed, it is known that the details of the regulation of nitrogen assimilation in higher plants do differ from those just discussed. Additional complexity is to be expected because glutamine synthetase in higher plants exists in multiple isoenzymic forms with differing intracellular distribution and in association with different tissues (for a general review, see Forde and Cullimore 1989). For example, it has been shown that the distinct genes coding for the different glutamine synthetase polypeptides in *Arabidopsis thaliana* respond differentially to environmental stimuli (Peterman and Goodman 1991).

It would be surprising, however, if the same patterns of control alluded to in chapter 4 were not found to apply to the major producers. The themes of allosteric modulation, cascadic covalent modification, and the repression and derepression of specific segments of the genome recur across all species. It has been suggested recently, on the basis of the identification of sequences homologous to the "two-component regulators" (Chang et al. 1993 for *A. thaliana*; Ota and Varshavsky 1993), that pathways similar to the NR_I/NR_{II} system (phosphorylation of an aspartate residue in a response regulator by a histidine phosphorylating autokinase) do occur in eucaryotes. Koshland (1993) comments on these findings: "It now seems possible that all of the procaryotic permutations [of the two-component pathway] may be found in eucaryotes." One might extrapolate from such comment to suggest that similar schemes may account for the transduction of nutrient level signals into the metabolic responses of eucaryotic cells. This remains to be demonstrated, and considerable interest attaches to finding particular instantiations of such schemes in the autotrophs that play quantitatively significant roles in the global biogeochemical cycles.

On a global scale, any such fine-tuning of glutamine synthetase activity (assuming that it is a widespread characteristic among the global biota) may provide a solution at the molecular level to the problem posed earlier concerning the correlation of assimilation rates in the cycles of two elements (in this case, carbon and nitrogen). The relative flows of carbon and nitrogen into biomass must be adjusted to give an appropriate final stoichiometry of biomass (Redfield ratio). But this is done, in effect, by adjusting the rate of the assimilation reaction (glutamine synthesis) at a given concentration of the limiting nutrient. Consider an ecological situation in which carbon is readily available but nitrogen is in short supply. A simple mass action approach would predict a biomass rich in carbon but poor in nitrogen. This does not happen. The biota do not adjust the composition of their active cellular material in accordance with nutrient supply.

How is the constancy of the Redfield ratio (C/N) maintained? It appears likely that a high C/N ratio in the external environment will, all other environmental variables being held constant, tend to a high ratio of intracellular 2-oxoglutarate/glutamine. The control logic that I have just outlined predicts, under these circumstances, a decrease in the extent of adenylylation of glutamine synthetase. Consequently, nitrogen is incorporated more actively into the starting materials for the synthesis of the nitrogen-containing biopolymers, proteins, and nucleic acids. Liebig's Laws continue to be observed, but the efficacy of the limiting nutrient is changed, either by changing the constants governing the catalysis of assimilation or by changing the amount of catalytic protein. For the enteric bacteria we know that the former is brought about by the adenylylation/deadenylylation of glutamine synthetase and the latter by activation of the *gln* operon by NR_I-P. In each case, the metabolism of carbon and of nitrogen at the ecosystem level are seen to be connected through the molecular regulation of intracellular catalysis.

The Molecular Control of Nitrate Reductase

The enzymically catalyzed incorporation of more than 1 Patom/y of nitrogen into organic form implies, for a steady-state, that ammonia (NH_3) must be provided to the global stock of glutamine synthetase at the same rate. If ammonia nitrogen is assimilated into nitrogenous compounds of biomass, then the nutrient NH_3 must be regenerated from somewhere. Some of this NH_3 will be made available to primary producers by the decomposition of biomass and bioproduct (which includes excreta). But there is also an important anthropogenic role, as much of the nitrogen applied to agricultural land is in this reduced form of nitrogen.

Active microbial oxidation will then transform much of this NH_3 into nitrate. Nitrate is the principal form of nitrogen available to plants. But, as I have pointed out in the discussion of figure 6.3, NO_3^-, although ecologically important as a plant nutrient, is not converted directly to the complex organic nitrogen of biomass. It must first be reduced to NH_3. The reduction of NO_3^- to NH_3 must therefore be quantitatively second in importance among the enzymic reactions of global nitrogen metabolism. Not only does this reduction takes place as part of the assimilation of nitrogen but for many bacterial species the reduction of NO_3^- is an alternative form of respiration—NO_3^- taking the place of oxygen as terminal acceptor of reducing equivalents.

The enzymes involved in the formation of biomass are referred to as *assimilatory nitrate reductases*, while those involved in the specialized process of "nitrate respiration" are *dissimilatory nitrate reductases*. Globally, it would appear probable that assimilatory nitrate reduction must be quantitatively much more important than the dissimilatory process. In both cases the reaction is brought about in two steps. First is the 2-equivalent reduction of NO_3^- to NO_2^-. Second is the 6-equivalent reduction of NO_2^- to NH_3. This second step is catalyzed by nitrite reductase. The first of these two steps is probably the step that is subject to control (Guerrero et al. 1981). The level of activity of nitrite reductase is generally considerably greater than that of nitrate reductase in the same cells or tissues, and thus nitrite accumulation is not observed. Overall, the initial reduction of NO_3^- to NO_2^- appears to be the rate-limiting reaction, and so regulation would be expected here.

As in the case of glutamine synthetase, nitrate reductase is subject to control both at the gene level and by effects exerted on the activity of the enzyme. The reduction of NO_3^- to NH_3 consumes reducing equivalents that could be used in other important cellular biochemical processes. Guerrero, Vega, and Losada (1981) point out that the provision of nitrogen for the synthesis of biomass by the reduction of NO_3^- may require as much as 25% of the reducing equivalents needed in the corresponding reduction of CO_2. It would therefore be biologically advantageous to switch off the reduction of NO_3^- whenever NH_3 is available. A number of molecular mechanisms have been reported to achieve this end. One might expect that there would be a direct end-product feedback inhibition by NH_3, but such has not been observed. However, it does appear that the nitrate reductase of algae can exist in both an active form and an inactive form and that the conversion of active nitrate reductase to the inactive form is promoted by NH_3. In *Chlamydomonas* this inactivation appears to involve a change in the redox state of the enzyme, the reduced form being inactive (Guerrero et al. 1981), whereas in *Chlorella* inactivation occurs by a reaction of the reduced enzyme with CN^-. The active form of the enzyme can be regenerated in vitro by oxidation, concomitant with the release of stoichiometric amounts of CN^- (Lorimer et al. 1974). Nitrate reductase is a complex enzyme, containing three prosthetic groups—FAD, heme, and molybdopterin—so the inhibition by and binding of CN^- is not surprising. On the other hand, it is not clear where the CN^- is coming from, as no metabolic origin is known and thus the model for metabolic regulation of nitrate reductase by CN^- (Solomonson and Spehar 1979) remains speculative (Dunn-Coleman et al. 1984).

Most of what is known about the genetic regulation of nitrate reductase has been learned from the ascomycete fungi. But, as in the previous discussion of the regulation of glutamine synthetase in enteric bacteria, the details are presented here merely to exemplify the principles that are likely to govern the regulation of biochemical processes that have global significance.

The protein products of the genes responsible for the nitrate assimilation pathway (e.g., the *nit* genes of *Neurospora crassa*) are not so well characterized as those of the *gln* genes of the enteric bacteria. Induction of nitrate reductase by NO_3^- involves a regulatory gene (*nit* 4/5). Mutations in this gene behave recessively in heterozygous diploids, that is, only one copy of the unmutated gene is needed to regulate nitrate reductase. It has therefore been suggested that the gene product of *nit* 4/5 must exert a positive effect on the structural gene for nitrate reductase (*nit* 3). Certain mutants in this structural gene and some other associated genes are known to produce aberrant forms of nitrate reductase and to do so constitutively, that is, in the absence of NO_3^-.

On the basis of these findings, Cove and Pateman (1969) suggested a model for autologous regulation of nitrate reductase (figure 6.10). In such an autologous model, nitrate reductase activates the structural gene (*nit* 3) responsible for its own production. Such an activation is elicited, however, only in the presence of NO_3^-, when the process of NO_3^- reduction is advantageous. Activation in the absence of NO_3^- can occur either when the nitrate reductase is present in an aberrant form or when the organism is homozygous for defective forms of the regulatory gene (*nit* 4/5). The model of figure 6.10 indicates the gene product of *nit* 4/5 interacting, in the absence of NO_3^-, with nitrate reductase to form a complex that cannot activate *nit* 3. When NO_3^- is present, however, nitrate reductase cannot interact with the *nit* 4/5 product. Consequently, transcription of *nit* 3 is activated.

Organisms that are homozygous for mutants of *nit* 4/5 fail to produce the protein that binds to nitrate reductase, and aberrant forms of nitrate reductase (*nit* 3 mutants) fail to bind the regulatory product. In both types of mutants, the uncomplexed nitrate reductase will activate its own structural gene, and so enzyme production will take place without NO_3^- being present. The relevant regulatory genes (*nir* A in *Aspergillus nidulans*, *nit* 4 in *Neurospora crassa*) have been cloned. It should thus be possible to determine whether their protein products do indeed bind to nitrate reductase (Kinghorn 1989; Marzluf and Fu 1989; Scazzocchio and Arst 1989).

In addition to such activation of the nitrate reductase by the presence of substrate, production of the enzyme is repressed by nitrogenous products of nitrate assimilation such as glutamine. Considerable uncertainty surrounds the mechanism of this repression. The gene responsible for the repression, *nit* 2, as with *nit* 4/5, appears to code for a product that is essential for the activities of, among others, the structural gene *nit* 3. In *Aspergillus nidulans*, mutants in the *nit* 2 region have been isolated in which nitrate reductase production is not repressed by glutamine. Grove and Marzluf (1981) have isolated a protein capable of

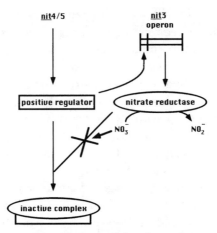

FIGURE 6.10 A possible scheme for autologous regulation of nitrate assimilation in *Neurospora crassa*. Production of nitrate reductase by *nit* 3 is turned on by the gene product of *nit* 4/5. The presence of NO_3^- prevents nitrate reductase from sequestering this positive regulator.
SOURCE: Adapted from figure 3 of Cove and Pateman 1969 and figure 1 of Dunn-Coleman, Smarelli, and Garrett 1984.

binding both DNA and glutamine, which may be the *nit* 2 product. Glutamine repression, on this view, would entail the binding of glutamine to this regulatory protein. in such a manner as to prevent its activation of *nit* 3 and thus to prevent the synthesis of nitrate reductase. An alternative view is that the inactivation of the *nit* 2 product is brought about by a product of the gene locus coding for glutamine synthetase (Dunn-Coleman et al. 1984). It has been assumed in such models that the repressor is glutamine synthetase itself, but it is not established that mutations—such as that of the strain of *A. nidulans* isolated by Mac-Donald (1982) which lacks glutamine synthetase and in which nitrate reductase is not repressed by glutamine—actually map within the structural gene for glutamine synthetase.

Either scheme (whether glutamine synthetase is directly involved or not) has similar biogeochemical implications. The possibility clearly exists for a molecular regulatory mechanism that integrates the activity of glutamine synthetase with the activity of nitrate reductase. On a global scale, glutamine synthetase catalyzes the assimilation of inorganic nitrogen into the organic nitrogen of biomass. It may well be of importance to global ecology that, through direct or indirect mechanisms, glutamine synthesis also regulates the activity of nitrate reductase, the enzyme that catalyzes the reduction of NO_3^- to provide the substrate for that key step in assimilation.

The assimilation of NO_3^- and that of NH_3 into biomass present different demands for reducing equivalents. Because the generation of reducing equivalents is tied in closely with the assembly and disassembly of the carbon skeletons of intermediary metabolites, assimilation of the two nutrient forms of nitrogen entails different interactions with carbon metabolism (Vanierberghe et al. 1992). It is important in this connection to note that in higher plants (specifically, *Medicago trunculata*) the multiple genes encoding for the isoenzymes of glutamine synthetase are expressed differentially in response to NH_3 and NO_3^- (Stanford et al. 1993). Once again it must be emphasized that the specific molecular mechanisms underlying ecological interrelationships must be those pertaining to those species that quantitatively dominate the biogeochemical flow in the ecosystem (global or local). This discussion of the details of the fungal *nit* genes is intended, rather, to exemplify the depth of analysis that will be needed to understand the control of global processes at the molecular level.

The Molecular Control of Nitrogenase

The third example of regulation at the molecular level that is amplified at the global level so as to regulate the biogeochemical cycle of nitrogen is the enzyme system nitrogenase. Nitrogenase plays a key role in the mobilization of nitrogen from its principal inorganic reservoir in the atmosphere (as N_2) into a nutrient form (NH_3) that can then be assimilated into biomass.

Although dinitrogen, N_2, is abundant (it constitutes 78% of the atmosphere), it is a relatively inert gas. Only a limited range of species among the procaryotes and archaebacteria are capable of reducing N_2 to NH_3. Globally, most of this nitrogen fixation into assimilable forms (10–20 Tatoms/y; see chapter 1) is carried out by symbiotic systems ($\approx$70%) and by cyanobacteria ($\approx$25%) (Dixon and Wheeler 1986). Most recent advances in understanding the enzymic reactions of N_2 fixation and their control have come from studies of the facultative anaerobe, *Klebsiella pneumoniae*.

It is likely that the features of the control of nitrogenase in this species

172 Molecular Regulation and Global Metabolism

are even more directly applicable to global pathways than can be said of the glutamine synthetase pathway described earlier in this chapter. This is because, for nitrogenase, there is no need to extrapolate to eucaryotic species, none of which is capable of fixing N_2. Many reviews of nitrogenase and its control are available (e.g., Phillips 1980, Orme-Johnson 1985, Ludden and Roberts 1990).

Two proteins are involved in the reduction of N_2 to NH_3. These are the MoFe protein (dinitrogenase), and the Fe protein (dinitrogenase reductase). Dinitrogenase is a tetramer made up of two nonidentical subunits, $\alpha_2\beta_2$. The α subunit is the product of the gene, *nif* K; the β subunit is the product of *nif* D. Dinitrogenase reductase is a dimer of identical subunits which are the products of *nif* H. The flow of reducing equivalents occurs as shown in figure 6.11. Two important features of the nitrogenase system, both related to the energy requirements for nitrogen fixation, are indicated in this figure. First, the reduction of N_2 is always accompanied by the reduction of H^+ to H_2. At least 25–30% of the reducing equivalents flowing through the system are "lost" in this manner (Simpson and Burris 1984), and the evolution of H_2 appears to be an obligatory step in the catalytic cycle of dinitrogenase. Second, there is a requirement for the free energy of hydrolysis of ATP. ATP is bound by dinitrogen reductase, and its hydrolysis accompanies the transfer of electrons from dinitrogenase reductase to dinitrogenase. The stoichiometry of the requirement is two ATP per electron transferred (Hageman et al. 1980).

The reduction of N_2 to NH_3 thus directly requires six reducing equivalents. And bear in mind that for every six electrons used for this reduction, another two electrons must be used for the reduction of H^+. The overall reaction may therefore be represented:

$$N_2 + 16ATP + 8e^- + 10H^+ \rightarrow 2NH_4^+ + H_2 + 16ADP + 16P_{inorg}$$

In the presence of an uptake hydrogenase system capable of utilizing H_2 and regenerating ATP, the energy requirement might drop from $16ATP/N_2$ down to 14. But it is clear that N_2 reduction is energetically expensive, and it is thus not surprising to find that the nitrogenase system is closely regulated.

As in the two cases previously discussed (molecular control of glutamine synthetase and of nitrate reductase), nitrogenase is regulated both by modification of already existing enzymes and at the gene level. The former effect is brought about by the ADP-ribosylation of

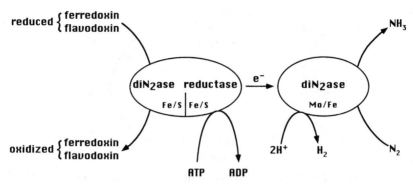

FIGURE 6.11 The flow of reducing equivalents and of energy in the nitrogenase complex.

dinitrogen reductase (Ludden and Roberts 1989). The genetic effect involves components of the system that activates the gene for glutamine synthetase.

Let us begin with the enzymic modification mode of nitrogenase regulation. The regulatory cycle is shown in figure 6.12. ADP-ribosyl groups are transferred from nicotinamide adenine dinucleotide (NAD) to an N-glycosidic bond with arginine, probably at position 100 of the dinitrogenase sequence. The transfer is catalyzed by an enzyme, dinitrogenase reductase ADP-ribosyl transferase (DRAT). The ADP-ribose group is removed by a different enzyme, dinitrogenase reductase-activating glycohydrolase (DRAG). This state of affairs contrasts with the cycles depicted in figures 6.6, 6.7, and 6.9, where both modification and removal of the modifying groups are catalytic functions of the same polypeptide. The effect of ADP-ribosylation is to inactivate dinitrogenase reductase; the enzyme is reactivated upon removal of the ADP-ribose. The modification (inactivation) takes place in cells in response to NH_3 and glutamine and, in photosynthetic organisms such as *Rhodospirillum rubrum*, when the cells are in the dark. Such a switch-off of the nitrogenase system in response to the supply of reduced nitrogen or in response to energy limitation has an obvious physiological rationale, but the ways in which these environmental signals affect the activities of DRAT and DRAG are not understood. The importance of cascade-like regulation is once again observed, since each molecule of DRAT and DRAG will be able to activate and inactivate many molecules of dinitrogenase reductase.

Let us now turn to the genetic means of regulating nitrogenase. The regulation of the nitrogenase system is quite complex. It occurs by way

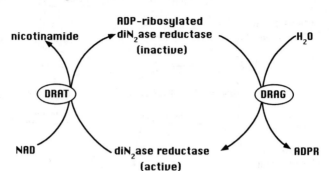

FIGURE 6.12 The cycle of ADP-ribosylation and de-ADP-ribosylation of dinitrogenase reductase.

of the *nif* genes. The primary role for the repression and derepression of nitrogenase is played by two genes, *nif* A and *nif* L, respectively. The gene product of the former (NIFA) activates the other *nif* genes. The transcription of *nif* A (and *nif* L) is activated, along with the *gln* genes, by the phosphorylated form of NR_I (which was introduced in the earlier section on glutamine synthetase). Thus the control logic,

$$\text{2-OG/GLN high} \therefore P_{II} \to P_{II}(UMP) \therefore NR_I \to NR_I\text{-P} \therefore gln \text{ active,}$$

must be extended for organisms capable of nitrogen fixation to include

$$\text{2-OG/GLN high} \therefore P_{II} \to P_{II}(UMP) \therefore NR_I \to NR_I\text{-P} \therefore nif \text{ A active} \therefore$$
$$\text{other } nif \text{ genes active.}$$

Thus, in such organisms, a low level of glutamine may turn on the genes required to fix N_2, thus providing NH_3 to the succeeding glutamine synthetase step. There is some suggestion that NIFA may resemble the response regulator for the *gln* operon NR_I (Ow and Ausubel 1983; Merrick 1983). Both the activation of the repressor gene *nif* L and the mode of action of the gene product, NIFL, are somewhat unclear. The expression of *nif* L requires NR_I-P, but other environmental factors are also needed. Among the signals that can bring about the expression of *nif* L are the presence of NH_3 and glutamine, although the former is inactive in strains lacking glutamine synthetase and therefore NH_3 as such cannot be the immediate signal for *nif* L expression. The repressor product, NIFL, appears to work by inactivating the activator product, NIFA (Arnott et al. 1989). Given the degree of homology between NIFA and NR_I (Drummond et al. 1986) it is natural to inquire whether the inactivation of NIFA by NIFL could be a dephosphorylation, analogous to the inactivation of NR_I by the phosphatase activity of NR_{II} (figure 6.9). Some evidence (Henderson et al. 1989) appears to indicate that this is not the case, that dephosphorylation is not involved in the inactivation of NIFA and that indeed the action of NIFL is not catalytic but, rather, involves a stoichiometric interaction between NIFA and NIFL.

Overall, the interaction of the *gln* operon and the *nif* operon provides a striking example of the complementary action of genes responsible for the coordination of two metabolic pathways that are both of biogeochemical significance. In addition, as discussed, a molecular basis for the interaction of glutamine metabolism and NO_3^- reduction probably exists, although this link is not as clearly established.

We have now explored three different means by which the global nitrogen cycle may be subject to molecular regulation. Each is mediated by a key enzyme: glutamine synthetase, nitrate reductase, and nitrogenase. These particular molecular feedback loops in the nitrogen cycle are, however, not the only ones that may have geochemical significance. There may be significant ecological roles for such effectors as cell membrane transporters for NO_3^- (Larsson and Ingemarsson 1989). Fossing et al. (1995) have provided a remarkable instance of the role of such molecular pumps in linking the cycles of nitrogen and sulfur in the sediments of the Western Pacific continental shelf. The filamentous bacterium, *Thioplaca*, is capable of concentrating NO_3^- from the seawater overlying the sediments to a remarkable degree (20,000×). Then, by downward migration of gliding filaments into the underlying sediment to zones rich in H_2S, the accumulated NO_3^- can be used as oxidant in energy-yielding oxidation-reduction reactions such as,

$$8NO_3^- + 5H_2S \rightarrow 4N_2 + 5SO_4^{2-} + 4H_2O + 2H^+$$

In commenting on these findings, Jannasch (1995) writes, "Other implications apart, the newly discovered transport system in the *Thioplaca* mats is likely to have a pronounced effect on the nitrogen cycle in the South American coastal upwelling zone."

Many other environmental factors could be thought of as regulating the nitrogen cycle of various ecosystems. Redox potential is an example that springs readily to mind. Nitrogenase itself is highly sensitive to O_2, and the expression of the nitrogen fixation genes is also controlled by O_2. Gilles-Gonzalez, Ditta, and Helinski (1991) have shown that the sensor for O_2 is an unusual hemoprotein (FixL) which can phosphorylate the regulatory protein FixJ (probably by modifying an aspartate residue). This phosphorylation is analogous to that of NR_I, as catalyzed by NR_{II}. There could be many other effectors, such as the *nod* genes (Long 1989; Fisher and Long 1992) which direct the production of the novel sulfated oligosaccharides that act as signal molecules in legume-*Rhizobium* associations (Lerouge et al. 1990; Schwedock and Long 1990; Truchet et al. 1991; Spaink et al. 1991).

Toward a Molecular Biology of Gaia

In one of their earlier papers on Gaia, James Lovelock and Lynn Margulis (1974) pose a question that is central to the consideration of Gaia as an empirically supportable concept: "What are the sensors, amplifiers and control mechanisms operating to maintain constant the steady-state chemical composition of the atmosphere?" A possible answer, suggested by even a cursory glance at the enzymes of nitrogen metabolism, is this: *The mechanisms responsible for global homeostasis are the same molecular mechanisms that subserve biological adaptation.* The biochemical mechanisms that, on the one hand, make possible adaptation to the planetary environment may, on the other hand, stabilize flow in the biogeochemical cycles.

It is perhaps no longer a particularly controversial idea that biochemical processes are involved in the feedback loops that have stabilized key characteristics of the planetary environment on a geological timescale. At the San Diego conference I referred to such a claim for "homeostasis *by* the biosphere" as the weak form of the Gaia hypothesis, as distinct from the strong form, "homeostasis *for* the biosphere" (Williams 1991). But it may be useful to make a further distinction between two ways in which biochemical processes could be involved in the feedback mechanisms of global environmental cycles. Figure 6.13 is an attempt to schematize the two modes and to clarify the distinction.

In both schemes depicted in this figure some subset of the global biota catalyzes flow through a biogeochemical cycle. In both cases the rate of flow in the cycle and the steady-state level of the environmental component will be influenced by the relevant kinetic characteristics of the biocatalyst. For the case of figure 6.13a, the response to the environmental component will be relatively simple, governed by the generalizations referred to earlier in this chapter as Liebig's Laws or by the kind of relationship between flux rate and the availability of nutrient indicated in

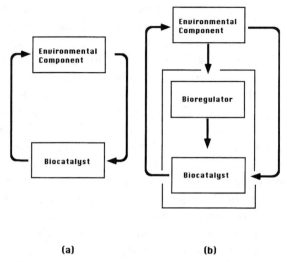

FIGURE 6.13 Two forms of biological involvement in an environmental cycle. In (a) the biotic component is catalytic but not necessarily regulatory. In (b) the sensitivity of the biological process to the environmental component gives it a regulatory role.

figure 6.1. In this mode, substrate limitation controls one step in a biogeochemical feedback loop. The case of figure 6.13b is more complex. It postulates a self-contained molecular regulatory mechanism that has dual significance—it is adaptive for the organisms that possess it, but it also regulates the overall rate of flow through a biogeochemical cycle.

It is the second of these schemes that gives depth to the metaphor of global metabolism. It implies that the biota cannot be treated as a "black box" as they are in the metaphor of a global chemical factory, traditionally used by geochemists (Garrels et al. 1973) and chemical oceanographers (Broecker and Peng 1982). Broecker and Peng in their text on chemical oceanography, *Tracers in the Sea*, entitle one of their chapters, "What keeps the system in whack?" We have seen in this final chapter that there is no shortage of biomolecular sensors and effectors in the nitrogen cycle that do have the potential to play a part in keeping the global nitrogen cycle "in whack." But if the question of Lovelock and Margulis (1974) concerning the "sensors, amplifiers, and control mechanisms" is to be answered in this way, then the "black box" will have to be opened. The fine details of molecular regulatory mechanisms elaborated through the evolutionary process must turn out to be determinative for the dynamics of global cycles.

Our understanding of the complicated cellular control mechanisms of nitrogen mobilization and assimilation, as presented even in cursory form in this chapter, demonstrates the level of analysis that will have to be achieved if global metabolism is to be fully understood. Molecular control systems such as these are conventionally thought of as adaptive mechanisms enabling organisms to occupy particular ecological niches. For instance, species possessing regulated *nif* genes will be better able to survive in habitats where fixed nitrogen fluctuates. The life boundary is expanded to include such habitats. But the regulation of *nif* genes can also be seen as serving to restrain environmental factors, such as ammonia concentrations, within limits that permit biotic survival.

We have seen in chapter 5 that one of the most telling criticisms of the Gaia hypothesis has concerned the difficulties of reconciling the idea of a global superorganism with evolutionary theory (Dawkins 1982). The selection of genes to bring about a metabolic pathway is a difficult problem, even within and for just a single species (for a particular case, see Petsko et al. 1993). Dawkins's solution to the problem (his Model 2) is to suggest that selection will favor a gene making a given enzyme because of the complementariness of that gene product to other products of the genome. For example, it may complete a metabolic chain or it may provide an adaptive regulatory mechanism enhancing survival. The difficulty in constructing a scenario for the evolution of a molecular mechanism for global homeostasis arises from the assumption that any such mechanism would have to be the product of the genomes of many organisms of many species. And the genes responsible for parts of the molecular machinery would have to improve their survivability by being complementary to other genes in very diverse organisms.

There is no theoretical objection to between-gene interactions occurring within populations rather than within organisms. Indeed, Dawkins's criticism of the Gaia hypothesis occurred within the context of a book in which the importance of such interactions was being stressed. Nonetheless, "Genes seem likely, other things being equal, to exert more power over nearby phenotypic characters than over distant ones. As an important special case of this, cells are likely to be quantitatively more heavily influenced by genes inside them than by genes inside other cells. The same will go for bodies" (Dawkins 1982:248).

What if, however, the basic assumption that underlies Dawkins's criticism is wrong? Perhaps the stability and integration of the biogeochemical cycles that suggests the Gaian metaphors is not brought about by *inter*genomic interactions. Is it possible that the accepted and relatively well understood *intra*genomic interactions that lead to the coordination of cellular metabolism may also have consequences at the ecological, even at the global, level? Here we can turn to the details of the metabolic and geochemical cycles of nitrogen for insight. A lesson may be drawn from the molecular regulation of glutamine synthetase, nitrate reductase, and nitrogenase. These three cases suggest the possibility that coordination of the global processes of assimilation, regeneration, and mobilization occurs *within* genomes.

Recall that the interaction of the global cycles of nitrogen and carbon takes place at the biological level. The internal metabolic processes that lead from the inorganic nutrients carbon dioxide and ammonia to biomass—assimilation—meet in the enzymic steps in which nitrogen is incorporated into glutamine. The rate of this incorporation is controlled by the activity of a key enzyme, glutamine synthetase. The formation of ammonia from nitrate—the final biochemical step in regeneration—is catalyzed by nitrate reductase. We have seen how the activity of this enzyme is regulated, and coordinated to that of glutamine synthetase, by the supply of fixed forms of nitrogen. Similarly, the activities of the enzymes responsible for the reduction of nitrogen to ammonia are finely tuned to the needs of the nitrogen-fixing microorganisms. The production of the key enzymes of the nitrogen cycle is closely coordinated at the genomic level, and the activities of these same enzymes are coordinated by allosteric effects and by cascades of enzymic modification.

It thus appears that the interaction at the biological level of the vast global biogeochemical cycles of carbon and nitrogen may be a reflection of events at the molecular level. One might assert that the regulation of the global biogeochemical cycles is the molecular control of intermediary metabolism writ large. From this perspective, global metabolism would be seen to be just as much a result of the properties of proteins synthesized under the control of genes as is cellular metabolism. These genes were selected, at least in part, because the complementarity of their products to other products in the same genome increased the efficiency of the cellular metabolism of those organisms that possess such a gene assembly. The global elemental cycles of the nutrient elements do indeed look like the metabolism of a quasi-organism, Gaia. But, perhaps the resemblance arises because these cycles are coordinated and regulated by the molecular mechanisms that underlie the metabolic processes of real organisms. Perhaps we should start to look at these mechanisms with their ecological, even global, significance in mind.

The title of this book is an invitation to biochemists and molecular biologists to adopt such a perspective. The idea of a molecular biology of Gaia will justify itself if it has heuristic value, if it suggests new avenues to be explored. It is not clear that such exploration, however, will lead to any substantiation of the Gaia hypothesis. It may fall far short of any such quasi-organismic integration of the global biogeochemical cycles. Such exploration will surely be essential to an understanding of the looser web that is implicit in the metaphor of global metabolism. Similarly, it is not clear that a more detailed knowledge of comparative metabolic regulation will suffice to establish the idea of global homeostasis; but it does appear certain that such knowledge is going to be necessary to answer the broader question posed by Peter Weyl in 1966: "Our problem is to discover the mechanisms by which the environmental variables are restricted within the life boundary."

This is not the place to debate the relative merits of holistic and reductionist approaches to biology. The holistic approach has an obvious appeal for ecologists and environmentalists, while biochemists and molecular biologists are, almost by definition, reductionists. Whatever the philosophical merits of the two positions, it cannot be denied that reductionism has turned out to be a supremely successful research strategy. In a work devoted to demolishing the metaphysical basis of reductionism, John Dupré (1993) concedes,

> If any recent scientific developments have tended to make reductionism look good, they are the spectacular advances in molecular genetics. The elucidation of the chemical structure of DNA and the subsequent elaboration, sometimes in exquisite detail, of some of the chemical processes involved in both reproduction and development are rightly counted among the major achievements of twentieth-century science.

By contrast, Dupré suggests that ecology is not a useful example from which to argue the failure of reductionism because it "suffers from the defect that it is not a part of science particularly likely to inspire reductionist enthusiasm."

Few ecologists would refrain from envy at the wealth of knowledge concerning organismic biology that has been generated by the reductionist strategy in the past half century. Many would echo the sentiments expressed by David Schindler and Everett Fee in 1975:

> The role of aquatic science today can perhaps be likened to that of medicine in the seventeenth century. Based upon the relatively com-

plete science of comparative anatomy, the purposes of most of the body's major organisms were precisely known, yet next to nothing was known about why the body functioned.

The solution to medical problems lay in the modern physiological approach to medicine, where a few compounds were found to exert fundamental chemical control over the functions of cells and organs of many sorts, greatly simplifying the understanding of the body. Examples are acetylcholine, adenosine triphosphate, nucleic acids and antigens. Progress was swift once these major changes in medical philosophy were made.

It appears that some equivalent drastic revision in approach to understanding the aquatic ecosystem may be necessary if solutions to a variety of environmental problems are to be obtained with confidence.

It is perhaps worth recalling that much of the success admired by Schindler and Fee has taken place since the solution to the genetic problem was given by Watson and Crick's elucidation of the double helical nature of DNA. It is relevant to recall that, under the influence of Schrödinger's writings, many in the field were expecting new physical principles to be invoked. As Francis Crick thought back on those heady days, "But then you ought to consider Max Delbrück. He went into it because he hoped that by looking at biological things you would find new laws of physics and chemistry" (quoted in Judson 1979:613).

Such hopes were not fulfilled. As Max Delbrück recalled, "the whole business was like a child's toy that you could buy at the dime store . . . all built in this wonderful way that you could explain in *Life* magazine so that really a five-year-old can understand what's going on. That there was so simple a trick behind it. This was the greatest surprise for everyone" (quoted in Judson 1979:60).

Perhaps there is a lesson here for ecologists. Schindler and Fee are calling for some new ordering principle in limnology. Some have greeted "Gaia" as a new ordering principle in global ecology. Perhaps no such principle is needed. If the workings of (a real or metaphorical) Gaia are ever to be understood, they will have to be uncovered in the same way that the workings of all organisms are being exposed—by analyzing them piece by piece and discovering what mechanisms make them tick.

This chapter's exploration of the regulation of the pathways of nitrogen metabolism is, I hope, a start. We have seen that some significant aspects of the integration of biogeochemical cycles on all scales up to and including the global scale may take place at the molecular level. The surprise of global metabolism may be that a common molecular machinery subserves the feedback processes both of cellular homeostasis and of the environmental stabilization that sustains Earth's habitability.

References

Adams, J. M., H. Faure, L. Faure-Denard, J. M. McGlade, and F. I. Woodward. 1990. Increases in terrestrial carbon storage from the last Glacial Maximum to the present. *Nature* 348:711–714.

Alvarez, L., W. Alvarez, F. Asaro, and H. V. Michel. 1980. Extraterrestrial cause for the Cretaceous-Tertiary extinction. *Science* 208:1095–1108.

Anderson, J. G., D. W. Toohey, and W. H. Brune. 1991. Free radicals within the Antarctic vortex: The role of CFCs in Antarctic ozone loss. *Science* 251:39–46.

Anfinsen, C. B. 1962. The tertiary structure of ribonuclease. In *Brookhaven Symposium in Biology: 15 Enzyme Models and Enzyme Structure*, pp. 184–197. Upton: Brookhaven National Laboratory.

Archer, D. and E. Maier-Reimer. 1994. Effect of deep-sea sedimentary calcite preservation on atmospheric CO_2 concentration. *Nature* 367:260–263.

Arnold, S. J. 1987. Genetic correlation and the evolution of physiology. In M. E. Feder, A. F. Bennett, W. W. Burggren, and R. B. Huey, eds., *New Directions in Ecological Physiology*, pp. 189–215. Cambridge: Cambridge University Press.

Arnott, M., C. Sidoti, S. Hill, and M. Merrick. 1989. Deletion analysis of the nitrogen fixation regulatory gene, *nifL*, of *Klebsiella pneumoniae*. *Arch. Microbiol.* 151:180–182.

Atjay, G. L., P. Ketner, and P. Duvigneaud. 1979. Terrestrial primary production and phytomass. In B. Bolin, E. T. Degens, S. Kempe, and P. Ketner, eds., *SCOPE 13: The Global Carbon Cycle*, pp. 129–181. Chichester: John Wiley.

Ayers, G. P. and J. L. Gras. 1991. Seasonal relationship between cloud condensation nuclei and aerosol methanesulphonate in marine air. *Nature* 353:834–835.

Ayers, G. P., J. P. Ivey, and R. W. Gillett. 1991. Coherence between seasonal cycles of dimethyl sulphide, methanesulphonate, and sulphate in marine air. *Nature* 349:404–406.

Bacastow, R. and C. D. Keeling. 1973. Atmospheric carbon dioxide and radiocarbon in the natural carbon cycle: II. Changes from A.D. 1700 to 2070 as deduced from a geochemical model. In G. M. Woodwell and E. V. Pecan, eds., *Brookhaven Symposium in Biology 24 Carbon and the Biosphere*, pp. 86–135. Springfield, Virginia: U.S. Atomic Energy Commission.

Baker, S. C., D. P. Kelly, and J. C. Murrell. 1991. Microbial degradation of methane sulphonic acid: A missing link in the biogeochemical sulphur cycle. *Nature* 350:627–628.

Barlow, C. and T. Volk. 1990. Open systems living in a closed biosphere: A new paradox for the Gaia debate. *BioSystems* 23:371–384.

———. 1992. Gaia and evolutionary biology. *BioScience* 42:686–693.

Barnola, J. M., D. Raynaud, Y. S. Korotkevitch, and C. Lorius. 1987. Vostok ice core provides 160,000-year record of atmospheric CO_2. *Nature* 329:319–321.

Barrow, J. D. and F. J. Tipler. 1986. *The Anthropic Cosmological Principle*. Oxford: Oxford University Press.

Bassham, J. A. 1977. Increased crop production through more controlled photosynthesis. *Science* 197:630–638.

Bates, T. S., R. J. Charlson, and R. H. Gammon. 1987. Evidence for the climatic role of marine biogenic sulphur. *Nature* 329:319–321.

Benton, M. J. 1985. Mass extinction among nonmarine tetrapods. *Nature* 316:811–814.

———. 1995. Diversification and extinction in the history of life. *Science* 268:52–58.

Berger, W. H. 1984. Short-term changes affecting atmosphere, oceans, and sediments during the Phanerozoic: Group report. In H. Holland and A. F. Trendall, eds., *Patterns of Change in Earth Evolution*, pp. 171–205. Berlin: Springer-Verlag.

Bernard, C. (Translated by H. E. Hoff, R. Guillemin, and L. Guillemin). 1974. *Lectures on the Phenomena of Life Common to Animals and Plants*. Springfield: Charles C. Thomas.

Berner, R. A. 1990. Atmospheric carbon dioxide levels over Phanerozoic time. *Science* 249:1382–1386.

———. 1991. Atmospheric oxygen, tectonics, and life. In S. H. Schneider and P. J. Boston, eds., *Scientists on Gaia*, pp. 161–166. Cambridge, Mass.: MIT Press.

Berner, R. A. and D. E. Canfield. 1989. A new model for atmospheric oxygen over Phanerozoic time. *Amer. J. Sci.* 289:333–361.

Berner, R. A. and G. P. Landis. 1988. Gas bubbles in fossil amber as possible indicators of the major gas composition of ancient air. *Science* 239:1406–1409.

Bolin, B. 1983. C, N, P, and S cycles: Major reservoirs and fluxes. The carbon cycle. In B. Bolin and R. B. Cook, eds., *SCOPE 21: The Major Biogeochemical Cycles and Their Interactions*, pp. 41–45. Chichester: John Wiley.

———. 1986. How much CO_2 will remain in the atmosphere? The carbon cycle and projections for the future. In B. Bolin, B. R. Döös, J. Jäger and R. A. Warrick, eds., *SCOPE 29: The Greenhouse Effect, Climatic Change, and Ecosystems*, pp. 93–155. Chichester: John Wiley.

Bolin, B., E. T. Degens, P. Duvigneaud, and S. Kempe. 1979. The global biogeochemical carbon cycle. In B. Bolin, E. T. Degens, S. Kempe, and P. Ketner, eds., *SCOPE 13: The Global Carbon Cycle*, pp. 1–56. Chichester: John Wiley.

Bolle, H.-J., W. Seiler, and B. Bolin. 1986. Other greenhouse gases and aerosols:

Assessing their role for atmospheric radiative transfer. In B. Bolin, B. R. Döös, J. Jäger, and R. A. Warrick, eds., *SCOPE 29: The Greenhouse Effect, Climatic Change, and Ecosystems*, pp. 93–155. Chichester: John Wiley.

Bond, G., H. Heinrich, W. S. Broecker, L. Labeyrie, J. McManus, J. Andrews, S. Huon, R. Jantschik, S. Clasen, C. Simet, K. Tedesco, G. Bonani, and S. Ivy. 1992. Evidence for massive discharges of icebergs into the North Atlantic ocean during the last glacial period. *Nature* 360:245–249.

Bormann, F. H. and G. E. Likens. 1979. *Patterns and Process in a Forested Ecosystem*. Berlin: Springer-Verlag.

Botkin, D. B. 1977. Forests, lakes, and the anthropogenic production of carbon dioxide. *Bioscience* 27:325–331.

———. 1990. *Discordant Harmonies: A New Ecology for the Twenty-First Century.* Oxford: Oxford University Press.

Bowie, J. U., J. F. Reidhaar-Olson, W. A. Lim, and R. T. Sauer. 1990. Deciphering the message in protein sequences: Tolerance to amino-acid substitutions. *Science* 247:1306–1310.

Boyle, E. A. 1990. Quaternary deepwater paleoceanography. *Science* 249:863–870.

Bramryd, T. 1979. The effects of man on the biogeochemical cycle of carbon in terrestrial ecosystems. In B. Bolin, E. T. Degens, S. Kempe, and P. Ketner, eds., *SCOPE 13: The Global Carbon Cycle*, pp. 183–218. Chichester: John Wiley.

Broecker, W. S. 1970a. A boundary condition on the evolution of atmospheric oxygen. *J. Geophys. Res.* 75:3553–3557.

———. 1970b. Man's oxygen reserves. *Science* 168:1537–1578.

———. 1971. A kinetic model for the chemical composition of sea water. *Quaternary Res.* 1:188–207.

———. 1973. Factors controlling CO_2 content in the oceans and atmosphere. In G. M. Woodwell and E. V. Pecan, eds., *Brookhaven Symposium in Biology 24: Carbon and the Biosphere*, pp. 32–49. Springfield, Virginia: U.S. Atomic Energy Commission.

———. 1982. Ocean chemistry during glacial time. *Geochim. Cosmochim. Acta* 46:1689–1705.

———. 1987. Unpleasant surprises in the greenhouse. *Nature* 328:123–126.

———. 1994. Massive iceberg discharges as triggers for global climate change. *Nature* 372:421–424.

Broecker, W. S. and G. H. Denton. 1989. The role of ocean-atmosphere reorganizations in glacial cycles. *Geochim. Cosmochim. Acta* 53:2465–2501.

Broecker, W. S. and T.-H. Peng. 1982. *Tracers in the Sea*. Palisades, New York: Lamont-Doherty Geological Observatory.

———. 1992. Interhemispheric transport of carbon dioxide by ocean circulation. *Nature* 356:587–589.

Broecker, W. S., T. Takahashi, H. J. Simpson, and T.-H. Peng. 1979. Fate of fossil fuel carbon dioxide and the global carbon budget. *Science* 206:409–418.

Bunnell, F. L. and K. A. Scoullar. 1975. ABISKO II: A computer model of carbon flux in tundra ecosystems. In T. Rosswall and O. W. Heal, eds., *Ecol. Bull. 20: Structure and Function of Tundra Ecosystems*, pp. 425–448. Stockholm: Swedish Natural Research Council.

Burton, A. C. 1939. The properties of the steady state compared to those of equilibrium as shown in characteristic biological behaviour. *J. Cell. Comp. Physiol.* 14:327–349.

Caban, C. E. and A. Ginsburg. 1976. Glutamine synthetase adenylyltransferase from *Escherichia coli*: Purification and physical and chemical properties. *Biochemistry* 15:1569–1580.

Caemmerer, S. von and G. D. Farquhar. 1981. Some relationships between the biochemistry of photosynthesis and the gas exchange of leaves. *Planta* 153: 376–387.

Caldeira, K. and J. F. Kasting. 1992. The life span of the biosphere revisited. *Nature* 360:721–723.

Cerling, T. E. 1989. Does the gas content of amber reveal the composition of palaeoatmospheres? *Nature* 339:695–696.

Cess, R. D., G. L. Potter, J. P. Blanchet, G. J. Boer, S. J. Ghan, J. T. Kiehl, H. Le Treut, Z.-X. Li, X.-Z. Liang, J. F. B. Mitchell, J.-J. Morcrette, D. A. Randall,M. R.Riches, E. Roeckner, U. Schlese, A. Slingo, K. E. Taylor, W. M. Washington, R. T. Wetherald, and I. Yagai. 1989. Interpretation of cloud-climate feedback as produced by 14 atmospheric general circulation models. *Science* 245:513–516.

Challenger, F. 1959. *Aspects of the Organic Chemistry of Sulphur.* New York: Academic Press.

Chang, C., S. F. Kwok, A. B. Bleecker, and E. M. Meyerowitz. 1993. *Arabidopsis* ethylene-response gene *ETR1*: Similarity of product to two-component regulators. *Science* 262:539–544.

Chappell, J. and N. J. Shackleton. 1986. Oxygen isotopes and sea level. *Nature* 324:137–140.

Chappelaz, J., J. M. Barnola, D. Raynaud, Y. S. Korotkevitch, and C. Lorius. 1990. Ice-core record of atmospheric methane over the past 160,000 years. *Nature* 345:127–131.

Charles-Edwards, D. A. 1981. *The Mathematics of Photosynthesis and Productivity.* New York: Academic Press.

Charlson, R. J., J. E. Lovelock, M. O. Andreae, and S. G. Warren. 1987. Oceanic phytoplankton, atmospheric sulphur, cloud albedo, and climate. *Nature* 326: 655–661.

———. 1989. Correspondence. *Nature* 340:437–438.

Charlson, R. J. and T. M. L. Wigley. 1994. Sulfate aerosol and climatic change. *Sci. Amer.* 270:48–57.

Chisholm, S. W. and R. G. Stross. 1976. Phosphate uptake kinetics in *Euglena gracilis* (z) (Euglenophyceae) grown on light/dark cycles: I. Synchronized batch cultures. *J. Phycol.* 12:210–217.

Ciais, P., P. P. Tans, M. Trolier, J. W. C. White, and R. J. Francey. 1995. A large Northern Hemisphere terrestrial CO_2 sink indicated by the $^{13}C/^{12}C$ ratio of atmospheric CO_2. *Science* 269: 1098–1102.

Cofer, W. R. III, J. S. Levine, E. L. Winstead, and B. J. Stocks. 1991. New estimates of nitrous oxide emissions from biomass burning. *Nature* 349:689–691.

Cole, D. R. and H. C. Monger. 1994. Influence of atmospheric CO_2 on the decline of C4 plants during the last deglaciation. *Nature* 368:533–536.

Colombo, G. and J. J. Villafranca. 1986. Amino acid sequence of *Escherichia coli* glutamine synthetase deduced from the DNA nucleotide sequence. *J. Biol. Chem.* 261:10587–10591.

Conrad, R., W. Seiler, and G. Bunse. 1983. Factors affecting the loss of fertilizer nitrogen into the atmosphere as N_2O. *J. Geophys. Res.* 88:6709–6718.

Conway, T. J., P. P. Tans, L. S. Waterman, K. W. Thoning, D. R. Kitzis, K. A. Masarie, and N. Zhang. 1994. Evidence for interannual variability of the carbon cycle from the National Oceanic and Atmospheric Administration/Climate Monitoring and Diagnostics Laboratory Global Air Sampling Network. *J. Geophys. Res.*, Series D, 99:22831–22855.

Cove, D. J. and J. A. Pateman. 1969. Autoregulation of the synthesis of nitrate reductase in *Aspergillus nidulans*. *J. Bact.* 97:1374–1378.

Craig, H., C. C. Chou, J. A. Welham, C. M. Stevens, and A. Engelkemeir. 1988. The isotopic composition of methane in polar ice cores. *Science* 242:1535–1539.

Crutzen, P. J. 1983. Atmospheric interactions: Homogeneous gas reactions of C, N, and S containing compounds. 1983. In B. Bolin and R. B. Cook, eds., *SCOPE 21: The Major Biogeochemical Cycles and Their Interactions*, pp. 67–112. Chichester: John Wiley.

Crutzen, P. J. and M. O. Andreae. 1990. Biomass burning in the tropics: Impact on atmospheric chemistry and biogeochemical cycles. *Science* 250:1669–1678.

Crutzen, P. J. and F. Arnold. 1986. Nitric acid cloud formation in the cold Antarctic stratosphere: A major cause for the springtime "ozone hole." *Nature* 324:651–655.

Dacey, J. W. H. and S. G. Wakeham. 1986. Oceanic dimethylsulfide: Production during zooplankton grazing on phytoplankton. *Science* 233:1314–1316.

Dansgaard, W., J. W. C. White, and S. J. Johnsen. 1989. The abrupt termination of the Younger Dryas climate event. *Nature* 339:532–534.

Dawkins, R. 1982. *The Extended Phenotype.* Oxford: Oxford University Press.

——. 1986. *The Blind Watchmaker.* Harlow: Longman Scientific and Technical.

Degens, E. T. 1989. *Perspectives on Biogeochemistry.* Berlin: Springer-Verlag.

De Heer, J. 1957. The principle of Le Chatelier and Braun. *J. Chem. Ed.* 34:375–380.

——. 1958. Le Chatelier, scientific principle or "sacred cow." *J. Chem. Ed.* 35:133–136.

Delwiche, C. C. and G. E. Likens. 1977. Biological response to fossil fuel combustion products. In W. Stumm, ed., *Global Chemical Cycles and Their Alterations by Man*, pp. 73–88. Berlin: Dahlem Konferenzen.

Denton G. H. and C. H. Hendy. 1994. Younger Dryas advance of Franz Josef glacier in the southern Alps of New Zealand. *Science* 264:1434–1437.

Devol, A. H. 1991. Direct measurement of nitrogen gas fluxes from continental shelf sediments. *Nature* 349:319–321.

De Vooys, C. G. N. 1979. Primary production in aquatic environments. In B. Bolin, E. T. Degens, S. Kempe, and P. Ketner, eds., *SCOPE 13: The Global Carbon Cycle*, pp. 259–292. Chichester: John Wiley.

Diaz, S., J. P. Grime, J. Harris, and E. McPherson. 1993. Evidence of a feedback mechanism limiting plant response to elevated carbon dioxide. *Nature* 364:616–617.

Dickson, D. M., R. G. Wyn Jones, and J. Davenport. 1980. Steady state osmotic adaptation in *Ulva lactuca*. *Planta* 150:158–165.
DiMichele, L., K. Paynter, and D. A. Powers. 1991. Evidence of lactate dehydrogenase-B allozyme effects in the teleost, *Fundulus heteroclitus*. *Science* 253:898–900.
Dixon, R. K., S. Brown, R. A. Houghton, A. M. Solomon, M. C. Trexler, and J. Wisniewski. 1994. Carbon pools and flux of global forest ecosystems. *Science* 263:185–190.
Dixon, R. O. D. and C. T. Wheeler. 1986. *Nitrogen Fixation in Plants*. London: Blackie.
Drummond, M., P. Whitty, and J. Wooton. 1986. Sequence and domain relationships of *ntrC* and *nifA* from *Klebsiella pneumoniae*: Homologies to other regulatory proteins. *EMBO J.* 5:441–447.
Ducet, G., F. Blasco, and R. Jeanjean. 1977. Regulation of phosphate transport in *Chlorella pyrenoidosa* and *Candida tropicalis*. In E. Marré and O. Ciferri, eds., *Regulation of Cell Membrane Activity in Plants*, pp. 55–62. Amsterdam: North Holland.
Dumas, M. 1841. On the chemical statics of organized beings. *Phil. Mag.* 19 (3): 337–347, 456–469.
Dunn-Coleman, N. S., J. Smarelli, Jr., and R. H. Garrett. 1984. Nitrate assimilation in eukaryotic cells. *Int. Rev. Cytol.* 92:1–50.
Dupré, J. 1993. *The Disorder of Things: Metaphysical Foundations of the Disunity of Science*. Cambridge, Mass.: Harvard University Press.
Editorial. 1990. Doctrinal fallacies of stewardship. *Nature* 344:179–180.
Editorial. 1993. Environmental protection or imperialism? *Nature* 363:657–658.
Ehleringer, J. R., R. F. Sage, L. B. Flanagan, and R. W. Pearcy. 1991. Climate change and the evolution of C4 photosynthesis. *Trends Ecol. Evol.* 6:95–99.
Ehrlich, P. 1991. Coevolution and its applicability to the Gaia hypothesis. In S. H. Schneider and P. J. Boston, eds., *Scientists on Gaia*, pp. 19–22. Cambridge, Mass.: MIT Press.
Ellis, R. J. and S. M. Hemmingsen. 1989. Molecular chaperones: Proteins essential for the biogenesis of some macromolecular structures. *Trends in Biochem. Sci.* 14:339–342.
Elrifi, I. R., J. J. Holmes, H. G. Weger, W. P. Mayo, and D. H. Turpin. 1988. RuBP limitation of photosynthetic carbon fixation during NH_3 assimilation. Interaction between photosynthesis, respiration, and ammonium assimilation in N-limited green algae. *Plant Physiol.* 87:395–401.
Erwin, D. H. 1993. *The Great Paleozoic Crisis*. New York: Columbia University Press.
———. 1994. The Permo-Triassic extinction. *Nature* 367:231–235.
Fahnestock, G. R. 1979. Carbon input to the atmosphere from forest fires. *Science* 204:209–210.
Falk, A. E. 1981. Purpose, feedback, and evolution. *Philosophy of Science* 48:198–217.
Feder, M. E. 1986. The analysis of physiological diversity: The prospects for pattern documentation and general questions in ecological physiology. In M. E.

Feder, A. F. Bennett, W. W. Burggren, and R. B. Huey, eds., *New Directions in Ecological Physiology*, pp. 38–70. Cambridge: Cambridge University Press.

Ferek, R. J., R. B. Chatfield, and M. O. Andreae. 1986. Vertical distribution of dimethylsulphide in the marine atmosphere. *Nature* 320:514–516.

Fisher, R. F. and S. R. Long. 1992. *Rhizobium*-plant signal exchange. *Nature* 357: 655–660.

Foley, J. A., J. E. Kutzbach, M. T. Coe, and S. Levis. 1994. Feedbacks between climate and boreal forests during the Holocene epoch. *Nature* 371:52–54.

Forde, B. G. and J. V. Cullimore. 1989. The molecular biology of glutamine synthetase in higher plants. *Oxford Surveys Plant Mol. Cell Biol.* 6:247–296.

Fossing, H., V. A. Gallardo, B. B. Jorgensen, M. Hüttel, L. P. Nielsen, H. Schulz, L.P. Canfield, S. Forster, R. N. Glud, J. K. Gundersen, J. Küver, N. B. Ramsing, A. Teske, B. Thamdrup, and O. Ulloa. 1995. Concentration and transport of nitrate by the mat-forming sulphur bacterium *Thioplaca*. *Nature* 374:713–715.

Francey, R. J., P. P. Tans, C. E. Allison, I. G. Enting, J. W. C. White, and M. Troller. 1995. Changes in oceanic and terrestrial carbon uptake since 1982. *Nature* 373:326–330.

Freyer, H. D. 1979. Atmospheric cycles of trace gases containing carbon. In B. Bolin, E. T. Degens, S. Kempe, and P. Ketner, eds., *SCOPE 13: The Global Carbon Cycle*, pp. 101–128. Chichester: John Wiley.

Freyer, H.-D. and N. Belacy. 1983. $^{13}C/^{12}C$ records in northern hemispheric trees during the past 500 years: Anthropogenic impact and climatic superpositions. *J. Geophys. Res.* 88C:6844–6852.

Garcia, E. and S. G. Rhee. 1983. Cascade control of *Escherichia coli* glutamine synthetase. Purification and properties of P_{II} uridylyltransferase and uridylyl-removing enzyme. *J. Biol. Chem.* 258:2246–2258.

Garrels, R. M. and A. Lerman. 1981. Phanerozoic cycles of sedimentary carbon and sulfur. *Proc. Nat. Acad. Sci. U.S.A.* 78:4652–4656.

Garrels, R. M., A. Lerman, and F. T. Mackenzie. 1976. Controls of atmospheric O_2 and CO_2: Past, present, and future. *Amer. Scientist* 64:306–315.

Garrels, R. M., F. T. Mackenzie, and C. Hunt. 1975. *Chemical Cycles and the Global Environment*. Los Altos: William Kaufmann.

Garrels, R. M. and E. A. Perry, Jr. 1974. Cycling of carbon, sulfur, and oxygen through geologic time. In E. D. Goldberg, ed., *The Sea, Vol. 5: Marine Chemistry*, pp. 303–336. New York: John Wiley.

Garwin, L. 1988. Of impacts and volcanoes. *Nature* 336:714–716.

Gates, D. M. 1975. Conclusions: The challenge of the future for biophysical ecology. In D. M. Gates and R. B. Schmerl, eds., *Perspectives of Biophysical Ecology*, pp. 589–596. Berlin: Springer-Verlag.

———. 1985. Global biospheric response to increasing atmospheric carbon dioxide concentrations. In B. R. Strain and J. D. Cure, eds., *Direct Effects of Increasing Carbon Dioxide on Vegetation*, pp. 171–184. Washington, D.C.: U.S. Department of Energy.

Gavin, J., G. Kukla, and T. Karl. 1989. Correspondence. *Nature* 340:438.

Ghan, S. J., J. E. Pener, and K. E. Taylor. 1989. Correspondence. *Nature* 340:438.

Gilles-Gonzalez, M. A., G. S. Ditta, and D. R. Helinski. 1991. A haemoprotein

with kinase activity encoded by the oxygen sensor of *Rhizobium meliloti*. *Nature* 350:170–172.
Goldberg, M. E. 1985. The second translation of the genetic message: Protein folding and assembly. *Trends in Biochem. Sci.* 10:388–391.
Goldstein, J. A. and M. S. Brown. 1990. Regulation of the mevalonate pathway. *Nature* 343:425–430.
Gorham, E., P. M. Vitousek, and W. A. Reiners. 1979. The regulation of chemical budgets over the course of terrestrial ecosystem succession. *Ann. Rev. Ecol. Syst.* 10:53–84.
Gould, S. J. 1977. *Ever Since Darwin: Reflections in Natural History.* New York: W. W. Norton.
Graham, J. B., R. Dudley, N. M. Aguilar, and C. Gans. 1995. Implications of the late Palaeozoic oxygen pulse for physiology and evolution. *Nature* 375:117–120.
Grove, G. and G. A. Marzluf. 1981. Identification of the product of the major regulatory gene of the nitrogen control circuit of *Neurospora crassa* as a nuclear DNA-binding protein. *J. Biol. Chem.* 256:463–470.
Guerrero, M. G., J. M. Vega, and M. Losada. 1981. The assimilatory nitrate-reducing system and its regulation. *Ann. Rev. Plant. Physiol.* 32:169–204.
Häder, D.-P., R. C. Worrest, H. D. Kumar, and R. C. Smith. 1995. Effects of increased solar ultraviolet radiation on aquatic ecosystems. *Ambio* 24:174–180.
Hageman, R. V., W. H. Orme-Johnson, and R. H. Burris. 1980. Role of magnesium adenosine 5'-triphosphate in the hydrogen evolution reaction catalyzed by nitrogenase from *Azotobacter vinelandii*. *Biochemistry* 19:2333–2342.
Hansen, J. E. and A. A. Lacis. 1990. Sun and dust versus greenhouse gases: An assessment of their relative roles in global climate change. *Nature* 346:13–719.
Harrison, K. G., W. S. Broecker, and G. Bonani. 1993. The effect of changing land use on soil radiocarbon. *Science* 262:725–726.
Harte, J. 1991. Ecosystem stability and diversity. In S. H. Schneider and P. J. Boston, eds., *Scientists on Gaia*, pp. 77–79. Cambridge, Mass.: MIT Press.
Hays, J. D., J. Imbrie, and N. J. Shackleton. 1976. Variations in the earth's orbit: Pacemaker of the ice ages. *Science* 194:1121–1132.
Heimann, M., C. D. Keeling, and I. Y. Fung. 1986. Simulating the atmospheric carbon dioxide distribution with a three-dimensional tracer model. In J. R. Trabalka and D. E. Reichle, eds., *The Changing Carbon Cycle: A Global Analysis*, pp. 16–49. Berlin: Springer-Verlag.
Heinrich, H. 1988. Origins and consequences of cyclic ice rafting in the northeast Atlantic Ocean during the past 130,000 years. *Quaternary Res.* 29:142–152.
Henderson, L. J. 1970. *The Fitness of the Environment.* Gloucester, Mass.: Peter Smith.
Henderson, N., S. Austin, and R. A. Dixon. 1989. Role of metal ions in negative regulation of nitrogen fixation by the *nifL* gene product from *Klebsiella pneumoniae*. *Mol. Gen. Genetics* 216:484–491.
Henderson-Sellers, A. and K. McGuffie. 1989. Correspondence. *Nature* 340:436–437.
Hewitt, C. N., G. L. Kok, and R. Fall. 1990. Hydroperoxides in plants exposed to ozone mediate air pollution damage to alkene emitters. *Nature* 344:56–58.

Hochachka, P. W. and G. N. Somero. 1984. *Biochemical Adaptation*. Princeton: Princeton University Press.
Holland, H. D. 1978. *The Chemistry of the Atmosphere and Oceans*. New York: John Wiley.
———. 1984. *The Chemical Evolution of the Atmosphere and Oceans*. Princeton: Princeton University Press.
———. 1990. Origins of breathable air. *Nature* 347:17.
———. 1991. The mechanisms that control the carbon dioxide and oxygen content of the atmosphere. In S. H. Schneider and P. J. Boston, eds., *Scientists on Gaia*, pp. 174–179. Cambridge, Mass.: MIT Press.
Holser, W. T., M. Schidlowski, F. T. Mackenzie, and J. B. Maynard. 1988. Geochemical cycles of carbon and sulfur. In C. B. Gregor, R. M. Garrels, F. T. Mackenzie, and J. B. Maynard, eds., *Chemical Cycles in the Evolution of the Earth*, pp. 105–173. New York: John Wiley.
Hough, A. M. and R. G. Derwent. 1990. Changes in the global concentration of tropospheric ozone due to human activities. *Nature*. 344:645–648.
Houghton, R. A. 1993. Is carbon accumulating in the northern temperate zone? *Global Biogeochem. Cycles* 7:611–617.
Hsü, K. J. 1984. Geochemical markers of impacts and their effects on environments. In H. D. Holland and A. F. Trendall, eds., *Patterns of Change in Earth Evolution*, pp. 63–74. Berlin: Springer-Verlag.
Hsü, K. J. and J. A. McKenzie. 1985. A 'Strangelove' ocean in the earliest Tertiary. In E. T. Sundquist and W. S. Broecker, eds., *The Carbon Cycle and Atmospheric CO2: Natural Variations Archean to Present*, pp. 487–492. Washington, D. C.: American Geophysical Union.
Hudson, R. J. M., S. A. Gherini, and R. A. Goldstein. 1994. Modeling the global carbon cycle: Nitrogen fertilization of the terrestrial biosphere and the "missing" CO_2 sink. *Global Biogeochem. Cycles* 8:307–333.
Hunt, H. W., J. W. B. Stewart, and C. V. Cole. 1983. A conceptual model for interactions among carbon, nitrogen, sulphur, and phosphorus in grasslands. In B. Bolin and R. B. Cook, eds., *SCOPE 21: The Major Biogeochemical Cycles and Their Interactions*, pp. 303–325. Chichester: John Wiley.
Hutton, J. 1788. Theory of the earth, or an investigation of the laws observable in the composition, dissolution, and restoration of land upon the globe. *Trans. Roy. Soc. Edinburgh* 1:209–304.
Ivanov, M. V. 1983. Major fluxes of the global biogeochemical cycle of sulphur. In M. V. Ivanov and J. R. Freney, eds., *SCOPE 19: The Global Biogeochemical Sulphur Cycle*, pp. 449–463. Chichester: John Wiley.
Jackson, D. C. 1987. Assigning priorities among interacting physiological systems. In M. E. Feder, A. F. Bennett, W. W. Burggren, and R. B. Huey, eds., *New Directions in Ecological Physiology*, pp. 310–326. Cambridge: Cambridge University Press.
Jacob, F. and J. Monod. 1961. Genetic regulatory mechanisms in the synthesis of proteins. *J. Mol. Biol.* 3:318–356.
Jannasch, H. W. 1995. Life at the sea floor. *Nature* 374:676–677.
Jenkinson, D. S., D. E. Adams, and A. Wild. 1991. Model estimates of CO_2 emissions from soil in response to global warming. *Nature* 351:304–306.

Johnson, K. R., D. J. Nichols, M. Atrep, Jr., and C. J. Orth. 1989. High-resolution leaf-fossil record spanning the Cretaceous/Tertiary boundary. *Nature* 340: 708–711.

Jones, A., D. L. Roberts, and A. Slingo. 1994. A climate model study of indirect radiative forcing by anthropogenic sulphate aerosols. *Nature* 370:450–453.

Jorgensen, B. B. 1983. The microbial sulphur cycle. In W. E. Krumbein, ed., *Microbial Geochemistry*, pp. 91–124. Oxford: Blackwell.

Jouzel, J., C. Lorius, J. R. Petit, N. I. Barkov, and V. M. Kotlyakov. 1994. Vostok isotopic temperature record. In T. A. Boben, D. P. Kaiser, R. J. Sepanski, and F. W. Stoss, eds., *Trends '93: A Compendium of Data on Global Change*, pp. 590–602. Oak Ridge, Tenn.: ORNL/CDIAC-65. Carbon Dioxide Information Analysis Center, Oak Ridge National Laboratory.

Judson, H. F. 1979. *The Eighth Day of Creation*. New York: Simon and Schuster.

Kasting, J. F., O. B. Toon, and J. B. Pollack. 1988. How climate evolved on the terrestrial planets. *Sci. Amer.* 258:90–97.

Kauffman, S. A. 1993. *The Origins of Order Self-Organization and Selection in Evolution*. Oxford: Oxford University Press.

———. 1995. *At Home in the Universe: The Search for the Laws of Self-Organization and Complexity*. Oxford: Oxford University Press.

Keeling, C. D. 1994. Global historical CO_2 emissions. In T. A. Boben, D. P. Kaiser, R. J. Sepanski, and F. W. Stoss, eds., *Trends '93: A Compendium of Data on Global Change*, pp. 501–504. Oak Ridge, Tenn.: ORNL/CDIAC-65. Carbon Dioxide Information Analysis Center, Oak Ridge National Laboratory.

Keeling, C. D. and T. P. Whorf. 1994. Atmospheric CO_2 records from sites in the SIO air sampling network. In T. A. Boben, D. P. Kaiser, R. J. Sepanski, and F.W. Stoss, eds., *Trends '93: A Compendium of Data on Global Change*, pp. 16–26. Oak Ridge, Tenn.: ORNL/CDIAC-65. Carbon Dioxide Information Analysis Center, Oak Ridge National Laboratory.

Keeling, C. D., T. P. Whorf, M. Wahlen, and J. van der Plicht. 1995. Interannual extremes in the rate of rise of atmospheric carbon dioxide since 1980. *Nature* 375:666–670.

Keeling, R. F. 1991. Mechanisms for stabilization and destabilization of a simple biosphere: Catastrophe on Daisyworld. In S. H. Schneider and P. J. Boston, eds., *Scientists on Gaia*, pp. 118–120. Cambridge, Mass.: MIT Press.

Keeling R. F. and S. R. Shertz. 1992. Seasonal and interannual variations in atmospheric oxygen and implications for the global carbon cycle. *Nature* 358: 723–727.

Keener, J. and S. Kustu. 1988. Protein kinase and phosphoprotein phosphatase activities of nitrogen regulatory proteins NTRB and NTRC of enteric bacteria: Roles of the conserved amino-terminal domain of NTRC. *Proc. Nat. Acad. Sci. U.S.A.* 85:4976–4980.

Kell, R. G., D. B. Montluçon, F. G. Prahl, and J. I. Hedges. 1994. Sorptive preservation of labile organic matter in marine sediments. *Nature* 370:549–552.

Kerr, J. B. and C. T. McElroy. 1993. Evidence for large upward trends of ultraviolet-B radiation linked to ozone depletion. *Science* 262:1032–1034.

Khalli, M. A. K. and R. A. Rasmussen. 1994. Global decrease in atmospheric carbon monoxide concentration. *Nature* 345:702–705.

Kiene, R. P. and T. S. Bates. 1990. Biological removal of dimethyl sulphide from sea-water. *Nature* 345:702–705.

Kinghorn, J. R. 1989. Genetic, biochemical, and structural organization of the *Aspergillus nidulans crnA-niiA-niaD* gene cluster. In J. L. Wray, and J. R. Kinghorn, eds., *Molecular and Genetic Aspects of Nitrate Assimilation*, pp. 69–87. Oxford: Oxford University Press.

Kirchner, J. W. 1989. The Gaia hypothesis: Can it be tested? *Rev. Geophys.* 27: 223–235.

——. 1990. Gaia metaphor unfalsifiable. *Nature* 345:470.

——. 1991. The Gaia hypotheses: Are they testable? Are they useful? In S. H. Schneider and P. J. Boston, eds., *Scientists on Gaia*, pp. 38–46. Cambridge, Mass.: MIT Press.

Kivelson, M. G. and G. Schubert. 1986. Atmospheres of the terrestrial planets. In M. J. Kivelson, ed., *The Solar System: Observations and Interpretations*, pp. 117–134. Englewood Cliffs: Prentice-Hall.

Koehn, R. H. 1987. The importance of genetics to physiological ecology. In M. E. Feder, A. F. Bennett, W. W. Burggren, and R. B. Huey, eds., *New Directions in Ecological Physiology*, pp. 170–185. Cambridge: Cambridge University Press.

Kolata, G. 1986. Trying to crack the second half of the genetic code. *Science* 233:1037–1039.

Koshland, D. E. Jr. 1993. The two-component pathway comes to eukaryotes. *Science* 262:532.

Krebs, H. A. and H. L. Kornberg. 1957. *Energy Transformations in Living Matter*. Berlin: Springer-Verlag.

Krumbein, W. E. and P. K. Swart. 1983. The microbial carbon cycle. In W. E. Krumbein, ed., *Microbial Geochemistry*, pp. 5–62. Oxford: Blackwell.

Kudrass, H. R., H. Erlenkeuser, R. Vollbrecht, and K. Weiss. 1991. Global nature of the Younger Dryas cooling effect inferred from oxygen isotope data from Sulu Sea cores. *Nature* 349:406–409.

Kuhlbusch, T. A., J. M. Lobert, P. J. Crutzen, and P. Warneck. 1991. Molecular nitrogen emissions from denitrification during biomass burning. *Nature* 351: 135–137.

Kump, L. R. and J. E. Lovelock. 1995. The geophysiology of climate. In A. Henderson-Sellers, ed., *Future Climates of the World: A Modeling Perspective*. Amsterdam: Elsevier.

Kurtén, B. 1972. *The Ice Age*. Toronto: Longmans.

Lapenis, A. and M. R. Rampino. 1993. Predicting Earth's lifespan. *Nature* 363:218.

Larsson, C.-M. and B. Ingemarsson. 1989. Molecular aspects of nitrate uptake in higher plants. In J. L. Wray and J. R. Kinghorn, eds., *Molecular and Genetic Aspects of Nitrate Assimilation*, pp. 3–14. Oxford: Oxford University Press.

Lasaga, A. C. 1980. The kinetic treatment of geochemical cycles. *Geochim. Cosmochim. Acta* 44:815–828.

——. 1984. Chemical kinetics of water-rock interactions. *J. Geophys. Res.* 89: 4009–4025.

Lashof, D. A. and D. R. Ahuja. 1990. Relative contributions of greenhouse emissions to global warming. *Nature* 344:529–531.

Lee, C. 1994. Kitty litter for carbon control. *Nature* 370:503–504.

Legrand, M. R., R. J. Delmas, and R. J. Charlson. 1988. Climate forcing implications from Vostok ice-core sulphate data. *Nature* 334:418–420.

Legrand, M. R., C. Feniet-Saigne, E. S. Saltzman, C. Germain, N. I. Barkov, and V. N. Petrov. 1991. Ice-core record of oceanic emissions of dimethylsulphide during the last climate cycle. *Nature* 350:144–146.

Lemon, E. R. 1977. The land's response to more carbon dioxide. In N. R. Andersen and A. Malahoff, eds., *The Fate of Fossil Fuel CO_2 in the Oceans*, pp. 97–130. New York: Plenum Press.

———. 1983. Interpretive summary. In E. R. Lemon, ed., *CO_2 and Plants The Response of Plants to Rising Levels of Atmospheric Carbon Dioxide*, pp. 1–5. Boulder: Westview Press.

Lerouge, P., P. Roche, C. Faucher, F. Maillet, G. Truchet, J. C. Promé, and J. Dearié. 1990. Symbiotic host-specificity of *Rhizobium meliloti* is determined by a sulphated and acylated glucosamine oligosaccharide signal. *Nature* 344:781–784.

Levins, R. and R. C. Lewontin. 1985. *The Dialectical Biologist*. Cambridge, Mass.: Harvard University Press.

Lewontin, R. C. 1972. Adaptation. *Sci. Amer.* 239:156–169.

Liebig, J. 1855. Principles of agricultural chemistry with special reference to the late researches made in England. In L. R. Pomeroy, ed., 1974, *Cycles of Essential Elements*, pp. 11–28. Stroudsburg, Pa.: Dowden, Hutchinson, and Ross.

Lieth, H. 1975. Primary productivity of the major vegetation units of the world. In H. Lieth and R. H. Whittaker, eds., *Primary Productivity of the Biosphere*, pp. 203–215. Berlin: Springer-Verlag.

Likens, G. E., F. H. Bormann, and N. M. Johnson. 1981. Interactions between major biogeochemical cycles in terrestrial ecosystems. In G. E. Likens, ed., *SCOPE 17: Some Perspectives of the Major Biogeochemical Cycles*, pp. 93–112. Chichester: John Wiley.

Long, S. R. 1989. *Rhizobium*-legume nodulations: Life together in the underground. *Cell* 56:203–214.

Lorimer, G. H., H. S. Gewitz, W. Völker, L. P. Solomonson, and B. Vennesland. 1974. The presence of bound cyanide in the naturally inactivated form of nitrate reductase of *Chlorella vulgaris*. *J. Biol. Chem.* 249:6074–6079.

Lorius, C., J. Jouzel, D. Raynaud, J. Hansen, and H. LeTreut. 1990. The ice-core record: Climate sensitivity and future greenhouse warming. *Nature* 347:139–145.

Lovelock, J. E. 1972. Gaia as seen through the atmosphere. *Atmospheric Environment* 6:579–580.

———. 1975. Thermodynamics and the recognition of alien biospheres. *Proc. Roy. Soc. London B.* 189:167–181.

———. 1979. *Gaia: A New Look at Life on Earth*. Oxford: Oxford University Press.

———. 1986. Geophysiology: A new look at earth science. *Bull. Amer. Met. Soc.* 67:392–397.

———. 1988. *The Ages of Gaia*. New York: Norton.

———. 1990. Hands up for the Gaia hypothesis. *Nature* 344:100–102.

———. 1991a. Geophysiology: The science of Gaia. In S. H. Schneider and P. J. Boston, eds., *Scientists on Gaia*, pp. 3–10. Cambridge, Mass.: MIT Press.

———. 1991b. *Healing Gaia*. London: Gaia Books.

———. 1992. A numerical model for biodiversity. *Phil. Trans. Roy. Soc. London B.* 338:383–391.

Lovelock, J. E. and J. P. Lodge, Jr., 1972. Oxygen in the contemporary atmosphere. *Atmospheric Environment* 6:575–578.

Lovelock, J. E., R. J. Maggs, and R. A. Rasmussen. 1972. Atmospheric dimethyl sulphide and the natural sulphur cycle. *Nature* 237:452–453.

Lovelock, J. E. and L. Margulis. 1974. Homeostatic tendencies of the earth's atmosphere. *Origins of Life* 5:93–103.

Lovelock, J. E. and M. Whitfield. 1982. Life span of the biosphere. *Nature* 296:561–563.

Ludden, P. W. and G. P. Roberts. 1989. Regulation of nitrogenase activity by reversible ADP ribosylation. *Curr. Topics in Cell Reg.* 30:23–56.

Luecke, H. and F. A. Quiocho. 1990. High specificity of a phosphate transport protein determined by hydrogen bonds. *Nature* 347:402–406.

Lyell, C. 1830. *Principles of Geology: Being an Inquiry How Far the Former Changes of the Earth's Surface are Referable to Causes Now in Operation.* London: John Murray.

MacDonald, D. W. 1982. A single mutation leads to loss of glutamine synthetase and relief of ammonium repression in *Aspergillus*. *Curr. Genetics* 6:203–208.

Machta, L. and E. Hughes. 1970. Atmospheric oxygen in 1967 to 1970. *Science* 168:1582–1584.

Margulis, L. 1990. Kingdom Animalia: The zoological malaise from a microbial perspective. *Amer. Zoologist* 30:861–875.

Margulis, L. and G. Hinkle. 1991. The biota and Gaia: 150 years of support for environmental sciences. In S. H. Schneider and P. J. Boston, eds., *Scientists on Gaia*, pp. 11–18. Cambridge, Mass.: MIT Press.

Marland, G., R. J. Andres, and T. A. Boden. 1994. Global, regional, and national CO_2 emissions. In T. A. Boben, D. P. Kaiser, R. J. Sepanski, and F. W. Stoss, eds., *Trends '93: A Compendium of Data on Global Change*, pp. 505–584. Oak Ridge, Tenn.: ORNL/CDIAC-65. Carbon Dioxide Information Analysis Center, Oak Ridge National Laboratory.

Martin, J., T. Langer, R. Boteva, A. Schramel, A. L. Horwich, and F.-U. Hartl. 1991. Chaperonin-mediated protein folding at the surface of groEL through a 'molten globule'-like intermediate. *Nature* 352:36–42.

Martin, J. H., R. M. Gordon, and S. E. Fitzwater. 1990. Iron in Antarctic waters. *Nature* 345:156–158.

Martin, J. H., K. H. Coale, K. S. Johnson, S. E. Fitzwater, R. M. Gordon, S. J. Tanner, C. N. Hunter, V. A. Elrod, J. L. Nowicki, T. L. Coley, R. T. Barber, S. Lindley, A. J. Watson, K. Van Scoy, C. S. Law, M. I. Liddicoat, R. Ling, T. Stanton, J. Stockel, C. Collins, A. Anderson, R. Bidigare, M. Ondrusek, M. Latasa, F. J. Millero, K. Lee, W. Yao, J. Z. Zhang, G. Friederich, C. Sakamoto, F. Chavez, K. Buck, Z. Kolber, R. Greene, P. Falkowski, S. W. Chisholm, F. Hoge, R. Swift, J. Yungel, S. Turner, P. Nightingale, A. Hatton, P. Liss, and N. W. Tindale. 1994. Testing the iron hypothesis in ecosystems of the equatorial Pacific Ocean. *Nature* 371:123–129.

Marzluf, G. A. and Y.-H. Fu. 1989. Genetics, regulation, and molecular studies of nitrate assimilation in *Neurospora crassa*. In J. Wray and J. R. Kinghorn, eds., *Molecular and Genetic Aspects of Nitrate Assimilation*, pp. 314–327. Oxford: Oxford University Press.

Mellilo, J. M., A. D. McGuire, D. W. Kicklighter, B. Moore III, C. J. Vorosmarty, and A. L. Schloss. 1993. Global climate change and terrestrial net primary production. *Nature* 363:234–240.

Merrick, M. J. 1983. Nitrogen control of the *nif* regulon in *Klebsiella pneumoniae*: Involvement of the *ntrA* gene and analogies between *ntrC* and *nifA*. *EMBO J.* 2:39–44.

Mitchell, J. F. B., T. C. Johns, J. M. Gregory, and S. F. B. Tett. 1995. Climate response to increasing levels of greenhouse gases and sulphate aerosols. *Nature* 376:501–504.

Mitchell, J. F. B., C. A. Senior, and W. J. Ingram. 1989. CO_2 and climate: A missing feedback? *Nature* 341:132–134.

Mix, A. C. and N. G. Pisias. 1988. Oxygen isotope analyses and deep-sea temperature changes: Implications for rates of oceanic mixing. *Nature* 331:249–251.

Monod, J. 1942. *La Croissance des Cultures Bacteriennes*. Paris: Herman.

———. 1971. *Chance and Necessity: An Essay on the Natural Philosophy of Modern Biology*. New York: Knopf.

Moody, J. B., T. R. Worsley, and P. R. Manoogian. 1981. Long-term phosphorus flux to deep-sea sediments. *J. Sed. Petrology* 51:307–312.

Morel, F. M. M., J. R. Reinfelder, S. B. Roberts, C. P. Chamberlain, J. G. Lee, and D. Yee. 1994. Zinc and carbon co-limitation of marine phytoplankton. *Nature* 369:740–742.

Morgan, M. E., J. D. Kingston, and B. D. Marino. 1994. Carbon isotopic evidence for the emergence of C4 plants in the Neogene from Pakistan and Kenya. *Nature* 367:162–165.

Mortlock, R. A., C. D. Charles, P. N. Froelich, M. A. Zibello, J. Satzman, J. D. Hays, and L. H. Burckle. 1991. Evidence for lower productivity in the Antarctic Ocean during the last glaciation. *Nature* 351:220–223.

Munson, R. D. and J. P. Doll. 1958. The economics of fertilizer use in crop production. *Adv. Agron.* 11:133–169.

Neftel, A., E. Moor, H. Oeschger, and B. Stauffer. 1985. Evidence from polar ice cores for the increase in atmospheric CO_2 in the past two centuries. *Nature* 315:45–47.

Newell, R. E., H. G. Reichle, Jr., and W. Seiler. Carbon monoxide and the burning earth. 1989. *Sci. Amer.* 261:82–88.

Newsholme, E. A. and B. Crabtree. 1976. Substrate cycles in metabolic regulation and in heat generation. In R. M. S. Smellie and J. F. Pennock, eds., *Biochemical Adaptation to Environmental Change*. London: The Biochemical Society.

Ninfa, A. and B. Magasanik. 1986. Covalent modification of the *glnG* product, NR_I, by the *glnL* product, NR_{II}, regulates the transcription of the *glnALG* operon in *Escherichia coli*. *Proc. Nat. Acad. Sci. U.S.A.* 83:5909–5913.

Novelli, P. C., K. A. Masarie, P. P. Tans, and P. M. Lang. 1994. Recent changes in atmospheric carbon monoxide. *Science* 263:1587–1590.

Oechel, W., S. Cowles, N. Grulke, S. J. Hastings, B. Lawrence, T. Prudhomme, G. Riechers, B. Strain, D. Tissue, and G. Vourlitis. 1994. Transient nature of CO_2 fertilization in Arctic tundra. *Nature* 371:500–503.

Oechel, W. and B. R. Strain. 1985. Native species responses to increased atmospheric carbon dioxide concentration. In B. R. Strain and J. D. Cure, eds., *Direct Effects of Increasing Carbon Dioxide on Vegetation*, pp. 117–154. Washington, D.C.: U.S. Department of Energy.

Orme-Johnson, W. H. 1985. Molecular basis of biological nitrogen fixation. *Ann. Rev. Biophys. Biophys. Chem.* 14:419–459.

Ota, I. M. and A. Varshavsky. 1993. A yeast protein similar to bacterial two-component regulators. *Science* 262:566–569.

Ow, D. W. and F. M. Ausubel. 1983. Regulation of nitrogen metabolism genes by *nifA* gene product in *Klebsiella pneumoniae*. *Nature* 301:307–313.

Palm, D. R., R. Goerl, and K. Burger. 1985. Evolution of catalytic and regulatory sites in phosphorylases. *Nature* 313:500–502.

Parikh, J. K. and J. P. Painuly. 1994. Population, consumption patterns, and climate change: A socioeconomic perspective from the South. *Ambio* 23: 434–437.

Pearcy, R. W. and O. Björkman. 1983. Physiological effects. In E. R. Lemon, ed., *CO2 and Plants: The Response of Plants to Rising Levels of Atmospheric Carbon Dioxide*, pp. 65–105. Boulder: Westview Press.

Peng. T.-H. and W. S. Broecker. 1984. Ocean life cycles and the atmospheric CO_2 content. *J. Geophys. Res.* 89:8170–8180.

Peng. T.-H., W. S. Broecker, H.-D. Freyer, and S. Trumbore. 1983. A deconvolution of the tree ring–based ^{13}C record. *J. Geophys. Res.* 88:3609–3620.

Perutz, M. F. 1989. Mechanisms of cooperativity and allosteric regulation in proteins. *Quart. Rev. Biophys.* 22:139–236.

Peterman, T. K. and H. M. Goodman. 1991. The glutamine synthetase gene family of *Arabidopsis thaliana*: Light-regulation and differential expression in leaves, roots, and seeds. *Mol. Gen. Genetics* 230:145–154.

Peterson, B. J. and J. M. Mellilo. 1985. The potential storage of carbon caused by eutrophication of the biosphere. *Tellus* 37B:117–127.

Petsko, G. A., G. L. Kenyon, J. A. Gerlt, D. Ringe, and J. W. Kozarich. 1993. On the origin of enzymatic species. *Trends in Biochem. Sci.* 18:372–376.

Phillips, D. A. 1980. Efficiency of symbiotic nitrogen fixation in legumes. *Ann. Rev. Plant Physiol.* 31:29–49.

Phillips, D. C. 1966. The three-dimensional structure of an enzyme molecule. *Sci. Amer.* 215:78–90.

Pomeroy, L. R. 1974. Editor's comments on papers 21 through 25. In L. R. Pomeroy, 1974, *Cycles of Essential Elements*, pp. 302–304. Stroudsburg, Penn.: Dowden, Hutchinson, and Ross.

Poorman, R. A., R. Randolph, R. Kemp, and R. Heinrikson. 1985. Evolution of phosphofructokinase: Gene duplication and creation of new effector sites. *Nature* 309:467–469.

Powers, D. A. 1987. A multidisciplinary approach to the study of genetic variation within species. In M. E. Feder, A. F. Bennett, W. W. Burggren, and R. B. Huey, eds., *New Directions in Ecological Physiology*, pp. 102–130. Cambridge: Cambridge University Press.

Prentice, K. C. and I. Y. Fung. 1990. The sensitivity of terrestrial carbon storage to climate change. *Nature* 346:48–51.

Proffitt, M. H., D. W. Fahey, K. K. Kelly, and A. F. Tuck. 1989. High-latitude ozone loss outside the Antarctic ozone hole. *Nature* 342:233–237.

Prospero, J. M., D. L. Savoie, E. S. Saltzman, and R. Larsen. 1991. Impact of oceanic sources of biogenic sulphur on sulphate aerosol concentrations at Mawson, Antarctica. *Nature* 350:221–223.

Ramanathan, V., R. D. Cess, E. F. Harrison, P. Minnis, B. R. Barkstrom, E. Ahmad, and D. Hartmann. 1989. Cloud-radiative forcing and climate: Results from the earth radiation budget experiment. *Science* 243:57–63.

Rampino, M. R. and T. Volk. 1988. Mass extinctions, atmospheric sulphur, and climatic warming at the K/T boundary. *Nature* 332:63–65.

Raup, D. M. 1984. Evolutionary radiations and extinctions. In H. D. Holland, and A. F. Trendall, eds., *Patterns of Change in Earth Evolution*, pp. 5–14. Berlin: Springer-Verlag.

Raup, D. M. and J. J. Sepkoski, Jr. 1982. Mass extinctions in the marine fossil record. *Science* 215:1501–1503.

Raval, A. and V. Ramanathan. 1989. Observational determination of the greenhouse effect. *Nature* 342:758–761.

Redfield, A. C. 1934. On the proportions of organic derivatives in sea water and their relation to the composition of plankton. *James Johnstone Memorial Volume*, pp. 176–192. Liverpool: Liverpool University Press.

———. 1958. The biological control of chemical factors in the environment. *Amer. Scientist* 46:205–221.

Rhee, S., P. B. Chock, and E. R. Stadtman. 1989. Regulation of *Escherichia coli* glutamine synthetase. *Adv. Enzymol.* 62:37–92.

Richards, F. A. 1957. Oxygen in the ocean. *Mem. Geol. Soc. Amer.* 67:185–238.

Richey, J. E. 1983. C, N, P, and S cycles, major reservoirs and fluxes: The phosphorus cycle. In B. Bolin and R. B. Cook, eds., *SCOPE 21: The Major Biogeochemical Cycles and Their Interactions*, pp. 51–56. Chichester: John Wiley.

Riebesell, U., D. A. Wolf-Gladrow, and V. Smetacek. Carbon dioxide limitation of marine phytoplankton growth rates. *Nature* 361:249–251.

Rosswall, T. 1983. C, N, P, and S cycles, major reservoirs and fluxes: The nitrogen cycle. In B. Bolin and R. B. Cook, eds., *SCOPE 21: The Major Biogeochemical Cycles and Their Interactions*, pp. 46–50. Chichester: John Wiley.

Rudolph, J. 1994. Anomalous methane. *Nature* 368:19–20.

Sagan, C., W. R. Thompson, R. Carlson, D. Gurnett, and C. Hord. 1993. A search for life on Earth from the Galileo spacecraft. *Nature* 365:715–721.

Saigne, C. and M. Legrand. Measurements of methanesulphonic acid in Antarctic ice. *Nature* 330:240–242.

Sarmiento, J. L. and E. T. Sundquist. Revised budget for the oceanic uptake of anthropogenic carbon dioxide. *Nature* 356:589–593.

Sathyendranath, S., T. Platt, E. P. W. Horne, W. G. Harrison, O. Ulloa, R. Outerbridge, and N. Hoepffner. 1991. Estimation of new production in the ocean by compound remote sensing. *Nature* 353:129–133.

Saunders, P. T. 1994. Evolution without natural selection: Further implications of the Daisyworld parable. *J. Theor. Biol.* 166:365–373.

Savageau, M. A. 1976. *Biochemical Systems Analysis*. Reading, Penn.: Addison-Wesley.

Savoie, D. L. and J. M. Prospero. 1989. Comparison of oceanic and continental sources of non-sea-salt sulphate over the Pacific ocean. *Nature* 339:685–687.

Scazzocchio, C. and H. N. Arst, Jr. 1989. Regulation of nitrate assimilation in *Aspergillus nidulans*. In J. L. Wray and J. R. Kinghorn, eds., *Molecular and Genetic Aspects of Nitrate Assimilation*, pp. 299–313. Oxford: Oxford University Press.

Schidlowski, M. 1991. Quantitative evolution of biomass through time: Biological and geochemical constraints. In S. H. Schneider and P. J. Boston, eds., *Scientists on Gaia*, pp. 211–222. Cambridge, Mass.: MIT Press.

Schidlowski, M., R. Eichmann, and C. E. Junge. 1975. Precambrian sedimentary carbonates: Carbon and oxygen isotope geochemistry and implication for the terrestrial oxygen budget. *Precamb. Res.* 2:1–69.

Schidlowski, M., J. M. Hayes, and I. R. Kaplan. 1983. Isotopic inferences of ancient biochemistries: Carbon, sulfur, hydrogen, and nitrogen. In J. W. Schopf, ed., *Earth's Earliest Biosphere: Its Origin and Evolution*, pp. 149–186. Princeton: Princeton University Press.

Schindler, D. W. and S. E. Bayley. 1993. The biosphere as an increasing sink for atmospheric carbon: Estimates from increased nitrogen deposition. *Global Biogeochem. Cycles* 7:717–733.

Schindler, D. W. and E. J. Fee. 1975. The roles of nutrient cycling and radiant energy in aquatic communities. In J. D. Cooper, ed., *Photosynthesis and Productivity in Different Environments*, pp. 323–343. Cambridge: Cambridge University Press.

Schlesinger, W. H. 1991. *Biogeochemistry: An Analysis of Global Change*. New York: Academic Press.

Schneider, S. H. and P. J. Boston. 1991. *Scientists on Gaia*. Cambridge, Mass.: MIT Press.

Schwartz, S. E. 1988. Are global cloud albedo and climate controlled by marine phytoplankton? *Nature* 336:441–445.

———. 1989. Correspondence. *Nature* 340:515–516.

Schwartzman, D. W. 1993. Comment on "Weathering, plants, and the long-term carbon cycle" by Robert A. Berner. *Geochim. Cosmochim. Acta* 57:2145–2146.

Schwartzman, D. W., S. N. Shore, T. Volk, and M. McMenamin. 1994. Self-organization of the Earth's biosphere—geochemical or geophysiological? *Origins Life Evol. Biosph.* 24:435–450.

Schwartzman, D. W. and T. Volk. 1989. Biotic enhancement of weathering and the habitability of Earth. *Nature* 340:457–460.

Schwedock, J. and S. R. Long. 1990. ATP sulphurylase activity of the *nodP* and *nodO* gene products of *Rhizobium meliloti*. *Nature* 348:644–647.

Seiler, W. and P. J. Crutzen. 1980. Estimates of gross and net fluxes of carbon between the biosphere and the atmosphere from biomass burning. *Clim. Change* 2:207–247.

Senior, P. J. 1975. Regulation of nitrogen metabolism in *Escherichia coli* and *Klebsiella aerogenes*: Studies with the continuous-culture technique. *J. Bact.* 123:407–418.

Shackleton, N. J. and N. G. Pisias. 1985. Atmospheric carbon dioxide, orbital forcing, and climate. In E. T. Sundquist and W. S. Broecker, eds., *The Carbon*

Cycle and Atmospheric CO_2: Natural Variations Archean to Present, pp. 303–317. Washington, D.C.: American Geophysical Union.
Shine, K. P., R. G. Derwent, D. J. Wuebles, and J.-J. Morcrette. 1990. Radiative forcing of climate. In J. T. Houghton, G. J. Jenkins, and J. J. Ephraums, eds., Climate Change: The IPCC Scientific Assessment, pp. 41–68. Cambridge: Cambridge University Press.
Siegenthaler, U. and J. L. Sarmiento. 1993. Atmospheric carbon dioxide and the ocean. Nature 365:119–125.
Sillén, L. G. 1961. The physical chemistry of seawater. In M. Sears, ed., Oceanography, pp. 549–581. Washington, D.C.: American Association for the Advancement of Science.
———. 1967. The ocean as a chemical system. Science 156:1189–1197.
Silverman, M. P. 1979. Biological and organic chemical decomposition of silicates. In P. A. Trudinger and D. J. Swaine, eds., Biogeochemical Cycling of Mineral-Forming Elements, pp. 445–465. Amsterdam: Elsevier.
Simpson, F. B. and R. H. Burris. 1984. A nitrogen pressure of 50 atmospheres does not prevent evolution of hydrogen by nitrogenase. Science 224:1095–1097.
Simpson, H. J. 1977. Man and the global nitrogen cycle: Group report. In W. Stumm, ed., Global Chemical Cycles and Their Alterations by Man, pp. 253–274. Berlin: Dahlem Konferenzen.
Slatyer, R. O. 1973. Energy flows in the biosphere. Proc. Linnean Soc. N. S. W. 97:225–236.
Slingo, A. 1990. Sensitivity of the Earth's radiation budget to changes in low clouds. Nature 342: 49–51.
Smith, W. O. Jr., L. A. Codispoti, D. M. Nelson, T. Manley, E. J. Buskey, H. J. Niebauer, and G. F. Cota. 1991. Importance of Phaeocystis blooms in the high-latitude ocean carbon cycle. Nature 352:514–516.
Söderlund, R. and B. H. Svensson. 1975. The global nitrogen cycle. In B. H. Svensson and R. Söderlund, eds., Ecol. Bull. 22 SCOPE 7: Nitrogen, Phosphorus, and Sulphur Global Cycles, pp. 23–73. Stockholm: Swedish Natural Research Council.
Solomon, S. 1988. The mystery of the Antarctic ozone "hole." Rev. Geophys. 26: 131–148.
Solomon, S. 1990. Progress towards a quantitative understanding of Antarctic ozone depletion. Nature 347–354.
Solomonson, L. P. and A. M. Spehar. 1979. Stimulation of cyanide formation by ADP and its possible role in the regulation of nitrate reductase. J. Biol. Chem. 254:2176–2179.
Spaink, H. P., D. M. Sheeley, A. A. M. van Brussel, J. Glushka, W. S. York, O. Geiger, E. P. Kennedy, V. N. Reinhold, and B. J. J. Lugtenburg. 1991. A novel highly unsaturated fatty acid moiety of lipo-oligosaccharide signals determines host specificity of Rhizobium. Nature 354:125–130.
Squyres, S. W. and J. F. Kasting. 1994. Early Mars: How warm and how wet? Science 265:744–749.
Stanford, A. C., K. Larsen, D. G. Barker, and J. V. Cullimore. 1993. Differential expression within the glutamine synthetase gene family of the model legume Medicago trunculata. Plant Physiol. 103:73–81.

Steel, D. 1995. *Rogue Asteroids and Doomsday Comets*. New York: Wiley.
Stevens, C. H. 1977. Was development of brackish oceans a factor in Permian extinctions? *Geol. Soc. Amer. Bull.* 88:133–138.
Stock, J. B., A. M. Stock, and J. M. Mottonen. 1990. Signal transduction in bacteria. *Nature* 344:395–400.
Stocker, T. F. and D. G. Wright. 1991. Rapid transitions of the ocean's deep circulation induced by changes in surface water fluxes. *Nature* 351:729–732.
Stuiver, M. 1978. Atmospheric carbon dioxide and carbon reservoir changes. *Science* 196:253–258.
Sundquist, E. T. 1993. The global carbon dioxide budget. *Science* 259:934–941.
Swisher, C. C. III, J. M. Grajales-Nishimura, A. Montanari, S. V. Margolis, P. Claeys, W. Alvarez, P. Renne, E. Cedillo-Pardo, F. J.-M. R. Maurrasse, G. H. Curtis, J. Smit, and M. O. McWilliams. 1992. Coeval ^{40}Ar/^{39}Ar ages of 65.0 million years ago from Chicxulub crater melt rock and Cretaceous-Tertiary boundary tektites. *Science* 257:954–957.
Tans, P. P., I. Y. Fung, and T. Takahashi. 1990 Observational constraints on the global atmospheric CO_2 budget. *Science* 247:1431–1438.
Todd, M. J., P. V. Viitanen, and G. H. Lorimer. 1994. Dynamics of the chaperonin ATPase cycle: Implications for facilitated protein folding. *Science* 265:659–666.
Tolba, M. K., O. A. El-Kholy, E. El-Hinnawi, M. W. Holdgate, D. F. McMichael, and R. E. Munn. 1992. *The World Environment 1972–1992: Two Decades of Challenge*. New York: Chapman and Hall.
Tolbert, N. E. and I. Zelitch. 1983. Carbon metabolism. In E. R. Lemon, ed., *CO2 and Plants: The Response of Plants to Rising Levels of Atmospheric Carbon Dioxide*, pp. 21–64. Boulder: Westview Press.
Toon, O. B. 1984. Sudden changes in atmospheric composition and climate. In H. D. Holland and A. F. Trendall, eds., *Patterns of Change in Earth Evolution*, pp. 41–61. Berlin: Springer-Verlag.
Truchet, G., P. Roche, P. Lerouge, J. Vasse, S. Camut, F. de Billy, J. C. Promé, and J. Dénarié. 1991. Sulphated lipo-oligosaccharide signals of *Rhizobium meliloti* elicit root nodule organogenesis in alfalfa. *Nature* 351:670–673.
Tschudy, R. H., C. L. Pillmore, C. J. Orth, J. S. Gilmore, and J. D. Knight. 1984. Disruption of the terrestrial plant ecosystem at the Cretaceous-Tertiary boundary, western interior. *Science* 225:1030–1032.
Turpin, D. H., F. C. Botha, R. G. Smith, R. Feil, A. K. Horsey, and G. C. Vanierberghe. 1990. Regulation of carbon partitioning to respiration during dark ammonium assimilation by the green alga *Selenastrum minutum*. *Plant Physiol.* 93:166–175.
Ullrich-Eberius, C. I., A. Noracky, E. Fischer, and U. Lüttge. Relationship between energy-dependent phosphate uptake and the electric membrane potential in *Lemna gibba* G1. *Plant Physiol.* 67:797–801.
Urey, H. C. 1952. *The Planets: Their Origin and Development*. New Haven: Yale University Press.
Vaghjiani, G. L. and A. R. Ravishankara. 1991. New measurement of the rate coefficient for the reaction of OH with methane. *Nature* 350:406–409.
Vairavamurthy, A., M. O. Andreae, and R. L. Iverson. 1985. Biosynthesis of

dimethylsulfide and dimethylpropiothetin by *Hymenomonas carterae* in relation to sulfur sources and salinity variations. *Limnol. Oceanogr.* 30:59–70.

Van Breemen, N., C. T. Driscoll, and J. Mulder. 1984. Acidic deposition and internal proton sources in acidification of soils and waters. *Nature* 307:439–443.

Van Cappellen, P. and E. D. Ingall. 1996. Redox stabilization of the atmosphere and oceans by phosphorus-limited marine productivity. *Science* 271:493–496.

Van der Hammen, T. 1988. Late-Tertiary and Pleistocene vegetation history, 20 My to 20 ky II.4 South America. In B. Huntley and T. Webb III, eds., *Vegetation History*, pp. 307–337. Dordrecht: Kluwer.

Vanierberghe, G. C., H. C. Huppe, K. D. M. Vlossak, and D. H. Turpin. 1992. Activation of respiration to support dark NO_3^- and NH_4^+ assimilation in the green alga *Selenastrum minutum*. *Plant Physiol.* 99:495–500.

Van Valen, L. 1971. The history and stability of atmospheric oxygen. *Science* 171:439–443.

Verniani, F. 1966. The total mass of the Earth's atmosphere. *J. Geophys. Res.* 71:385–391.

Vitousek, P. M. 1983. The effects of deforestation on air, soil, and water. In B. Bolin and R. B. Cook, eds., *SCOPE 21: The Major Biogeochemical Cycles and Their Interactions*, pp. 223–245. Chichester: John Wiley.

Waddington, C. H. 1957. *The Strategy of the Genes: A Discussion of Some Aspects of Theoretical Biology*. London: George Allen and Unwin.

Wahlen, M., N. Tanaka, R. Henry, B. Deck, J. Zeglen, J. S. Vogel, J. Southon, A. Shemesh, R. Fairbanks, and W. Broecker. 1990. Carbon-14 in methane sources and in atmospheric methane: The contribution from fossil carbon. *Science* 245:286–290.

Walker, J. C. G. 1977. *Evolution of the Atmosphere*. New York: Macmillan.

Walker, J. C. G., P. B. Hays, and J. F. Kasting. 1981. A negative feedback mechanism for the long-term stabilization of Earth's surface temperature. *J. Geophys. Res.* 86:9776–9782.

Walter, H. 1973. *Vegetation of the Earth*. Berlin: Springer-Verlag.

Ward, B. and R. Dubos. 1972. *Only One Earth*. Harmondsworth: Penguin Books.

Warrick, R. A., H. H. Shugart, M. Ya. Antonovsky, with J. R. Tarrant and C. J. Tucker. 1986. The effects of increased CO_2 and climatic change on terrestrial ecosystems: Global perspectives, aims, and issues. In B. Bolin, B. R. Döös, J. Jäger, and R. A. Warrick, eds., *SCOPE 29: The Greenhouse Effect, Climatic Change, and Ecosystems*, pp. 363–392. Chichester: John Wiley.

Watson, A. J. and J. E. Lovelock. 1983. Biological homeostasis of the global environment: The parable of Daisyworld. *Tellus* 35B:284–289.

Watson, A. J., C. Robinson, J. E. Robinson, P. J. le B. Williams, and M. J. R. Fasham. 1991. Spatial variability in the sink for atmospheric carbon dioxide in the North Atlantic. *Nature* 350:50–53.

Weatherley, A. H. 1990. Stewardship of resources. *Nature* 346:212.

Weiss, V. and B. Magasanik. 1988. Phosphorylation of nitrogen regulator I (NR_I) of *Escherichia coli*. *Proc. Nat. Acad. Sci. U.S.A.* 85: 8919–8923.

Weyl, P. K. 1966. Environmental stability of the earth's surface—chemical consideration. *Geochim. Cosmochim. Acta.* 30:663–679.

White, J. W. C., P. Ciais, R. A. Figge, R. Kenny, and V. Markgraf. 1994. A high-resolution record of atmospheric CO_2 content from carbon isotopes in peat. *Nature* 367:153–156.
Whittaker, R. H., G. E. Likens, and H. Lieth. 1975. Scope and purpose of this volume. In H. Lieth and R. H. Whittaker, eds., *Primary Productivity of the Biosphere*, pp. 3–5. Berlin: Springer-Verlag.
Wigley, T. M. L. 1989. Possible climate change due to SO_2-derived cloud condensation nuclei. *Nature* 339:365–367.
Wigley, T. M. L., R. Richels, and J. A. Edmonds. 1996. Economic and environmental choices in the stabilization of atmospheric CO_2 concentrations. *Nature* 379:240–243.
Williams, G. R. 1987. The coupling of biogeochemical cycles of nutrients. *Biogeochemistry* 4:61–75.
———. 1991. Gaian and nongaian explanations for the contemporary level of atmospheric oxygen. In S. H. Schneider and P. J. Boston, eds., *Scientists on Gaia*, pp. 167–173. Cambridge, Mass.: MIT Press.
Wilson, E. O. 1992. *The Diversity of Life*. Cambridge, Mass.: Belknap Press.
Wiman, I. M. B. 1990. Expecting the unexpected: Some ancient roots to current perceptions of nature. *Ambio* 19:62–69.
Wolfe, J. A. 1991. Palaeobotanical evidence for a June 'impact winter' at the Cretaceous/Tertiary boundary. *Nature* 352:420–423.
Wolfe, J. A. and G. R. Upchurch, Jr. 1986. Vegetation, climatic, and floral changes at the Cretaceous/Tertiary boundary. *Nature* 324:148–152.
Wong, C. S. 1978. Atmospheric input of carbon dioxide from burning wood. *Science* 200:197–200.
Woodwell, G. M., R. H. Whittaker, W. A. Reiners, G. E. Likens, C. C. Delwiche, and D. B. Botkin. 1978. The biota and the world carbon budget. *Science* 199:141–146.
Zachos, J. C., M. A. Arthur, and W. E. Dean. 1989. Geochemical evidence for suppression of pelagic marine productivity at the Cretaceous/Tertiary boundary. *Nature* 337:61–64.
Zelitch, I. 1971. *Photosynthesis, Photorespiration, and Plant Productivity*. New York: Academic Press.
Zepp, R. G., T. V. Callaghan, and D. J. Erikson. 1995. Effects of increased solar ultraviolet radiation on biogeochemical cycles. *Ambio* 24:181–187.
Zhang, R.-G., A. Joachimiak, C. L. Lawson, R. W. Schevitz, Z. Otwinowski, and P. B. Sigler. 1987. The crystal structure of *trp* aporepressor at 1.8 Å shows how binding tryptophan enhances DNA affinity. *Nature* 327: 591–597.

Index

Acid rain, 58
Adaptation: and the complexity of the environment, 137; biochemical strategies, 129–130; criticisms of the concept, 136; genetic origin of mechanisms for, 130; molecular link to natural selection, 130, 132; molecular mechanisms may also subserve global homeostasis, 176; molecular mechanisms of, 117, 129; time scales for, 133
Agricultural practices: application of nitrogenous fertilizers, 154, 156; as cause of increase in atmospheric N_2O, 54, 62; effects on N_2 fixation rate, 53; increase demand for phosphate, 55; linked to changes in atmospheric CO_2, 43
Anthropic principle, 69
Anthropogenic pressure: as a factor in post-glacial extinction, 13; greater effect on minor atmospheric pools, 64, 72; on negative feedback mechanisms, 113
Autocatalytic processes, 151

Biogeochemical cycles: and internal metabolic control mechanisms, 148, 177, 179; are they self-regulated?, 105; closedness, 17, 35; complex models, 17; dependence of global cycles on micro-cycles, 34; driven by photosynthesis, 24; fast sub-cycles isolated from slow cycles, 30, 34; four-box model, 17, 35, 38, 57, 148, 149, 152, 153; interaction in the metabolism of the biota, 35, 149, 152; interaction through enzyme regulation, 167, 180; linked to oceanic circulation, 30; non-linear relationships in, 150; stabilized by biochemical adaptation, 176, 177
Biomass, global: changes during glaciation, 13, 96; recent changes in, 43
Biota growth factor. *See* Michaelis constant
Bolide impacts: geochemical effects, 5; dust generated by, 6

C-4 plants: evolution of, 95
Carbon cycle: evidence for human impact on, 46; four-box model, 18, 48
Carbon dioxide: and solar luminosity, 77, 92; as greenhouse gas, 63; atmosphere-ocean exchanges, 39, 47; changes inferred from sedimentary ^{13}C, 12, 97; determined by silicate-carbonate equilibria, 74; effects of increase on plant growth, 43, 57, 92, 94; ice-core values at glacial-interglacial transitions, 12, 95, 97, 100; increase

due to fossil fuel combustion, 19, 41, 64; increase due to deforestation, 42; lower limits for photosynthesis, 78, 95; Mauna Loa records, 40; Michaelis constant of ribulose-1,5-bisphosphate carboxylase, 92, 153; missing sink, 47, 56, 94; rate of increase, 39; steady-state models, 76; trajectory from 3.8 Gya to present, 77, 92

Carbon isotopes: in tree rings, 44, 94

Carbon monoxide: sources and sinks, 51, 61

Carbon, organic: adsorbed on mineral surfaces, 85; ^{13}C content indicates biological origin, 65; burial on the ocean floor accompanied by increase in atmospheric O_2, 9, 29, 82–90; constancy of ratio to carbonate, 66

Carbonate-silicate relationships, 74, 79, 106

Carbonate ^{13}C, 66; changes negatively correlated to ^{34}S changes, 8

Carbonyl sulfide, 33, 57

Chemical equilibrium: as a cause of environmental stability, 69

Cloud condensation nuclei, 99

Clouds: effect on global radiation budget, 64; formation affected by anthropogenic sulfur emissions, 99; formation affected by biogenic sulfur emissions, 99

Daisyworlds, 138; no adaptation on, 140; optima in growth-temperature relationships, 141; stability built in by the modeler, 141; unstable variants, 139, 141

Deforestation: a cause of the increase in atmospheric CO_2, 42

Denitrification, 31, 54; estimates of global rate, 31, 156

Dimethyl sulfide, 33, 57, 98, 112; as a source of acid rain, 60; biochemical origin, 98; cloud-climate relationship, 99, 100; global input into the atmosphere, 98; high emissions in the last ice age, 100; oxidation resulting in the production of cloud condensation nuclei, 99; seasonal correlation with non-sea salt SO_4, 100

Dimethylsulfoniopropionate: as algal osmolyte, 99; as precursor of dimethyl sulfide, 98

Dust: generated by bolide impact, 6

Ecosystem: definition, 16

Environment, internal: atmosphere and hydrosphere analogous to, 108, 147; buffering as a biochemical strategy for adaptation, 130; constancy of, 104

Enzymes: allosteric modulation, 117, 121; cascades of, 118, 121, 160, 164, 173; feedback inhibition in systems of, 114; genetic regulation of, 119, 120, 121, 160, 175; isozymes and physiological function, 132; loss of regulation reduces fitness, 133; propensity to modulation is selected in evolutionary history, 113, 133; specificity of ligand binding, 118, 120, 121, 122, 127; specificity of modulation, 113

Eutrophication, 55, 58; global, 56, 97

Extinction: and limits to global stability, 7, 69; as a clue to changes in habitability, 4; correlation with geochemical markers, 5; greenhouse effect at the Cretaceous-Tertiary boundary, 97; multifactorial causation, 5; of megafauna at the end of the last glaciation, 13

Fertilization: global, 43, 56

Fire: in carbon cycle, 20, 27; in nitrogen cycle, 53, 55

Fitness of the environment, 69

Fossil fuel: annual rate of combustion, 41

Gaia, 98, 100, 101, 108; a product of luck?, 145; American Geophysical Union conference on, 108, 111, 137, 138, 141, 143, 176; as a challenge to evolutionary theory, 111; as a challenge to geochemical concepts, 111; criticized as teleological, 135; development requires genetic information, 144; difficulties concerning evolution of, 142, 145, 178; difficulties of accounting for gene cooperation in, 136; difficulty of integrating into mainstream biology, 123; homeorhetic development of, 144; homeostasis integral to, 136; identifies global biota as planetary control system, 111; molecular approach to, 181; optimality as a characteristic, 140; origins of the idea, 109; self-organization of, 144; unpopularity among scientists, 123

Genes: complementarity necessary for metabolic integration, 134, 136, 178; complementarity of the operons responsible for glutamine synthetase and nitrogenase, 175; inter-relationships between the genes for glutamine synthetase and for nitrate reductase, 170; operon model for regulation of, 120; transcription and translation of, 120

Geophysiology, 90, 102, 108, 109, 112, 113, 135, 147; *see also* Global metabolism

Glaciation, 9, 69, 95; effects on global biota, 13

Global metabolism: analogy to cellular metabolism, 104, 107; as a Gaian metaphor, 112, 122; as sum of metabolic activities of the global biota, 104; coordinated *within* genomes?, 179; involvement of biochemical processes in, 107; justification for the metaphor, 80, 90, 112; Mauna Loa records as icon of, 103; meaning, 103; regulation at the molecular level, 108, 177, 178; requires complementariness among genes of different species, 134, 178

Glutamate synthase, 158; reductant for, 158

Glutamine synthetase: 3-dimensional structure, 161; adenylylation stimulated by deuridylylated P_{II}, 161; allosteric effects on the transfer of adenylyl groups to and from, 162; deadenylylation requires uridylylated P_{II}, 161; effects of adenylylation of, 162; feedback inhibition of, 162; molecular link between carbon and nitrogen cycles, 158; multiple genes in higher plants expressed differentially in response to NH_3 and NO_3, 171; multiple isoenzymic forms in higher plants, 165; physiological effects of regulation of, 163; regulation of the gene for, 163; role in nitrogen cycle, 157; role of P_{II} in regulation of, 161; supply of NH_3 to, 167

Greenhouse effect: anthropogenic, 63; in glaciation-deglaciation, 12; in global temperature history, 77, 91

Gross primary production (GPP). *See* Production

Habitability: and adaptation, 123; as an ethical question, 2, 38; determined physico-chemically, 79, 146; end of, 2, 78; fragility of negative feedback mechanisms sustaining, 113; longevity of, 1, 65, 67, 143; not an all-or-none property, 125; role for the biota in, 146, 147; the "Goldilocks" theory, 70, 76

Heinrich events, 13

Hemoglobin: ecologically significant variations in, 133; model for allosteric modulation, 117

Homeostasis: *by* the biosphere or *for* the biosphere, 176; implied by persistence of life, 2, 67, 110; integral to Gaia, 136; physiological, 104, 105, 120, 147; physiological and global may share common molecular machinery, 182

Hydroxyl radical: as atmospheric oxidant, 51, 53, 54, 61, 99

Ice cores, 10; record of CH_4, 49; record of CO_2, 12, 44, 95; record of methyl-sulfonate, 100

Instability of the global environment, 2, 105; caused by biological activity, 100, 140; geological evidence for, 3

Isotopic shifts; in inorganic reservoirs, 7, 69; rapidity of the changes, 7

Le Chatelier's Principle, 75, 76, 79

Liebig's Law, 56, 149, 151, 176; reconciliation with Redfield ratios, 153, 166

Life: detected by non-equilibrium in the atmosphere, 109; geochemical evidence for persistence of, 87, 154

Life boundary, 68, 69, 72, 75, 76, 77, 95, 105, 112, 117, 123, 124, 125, 134, 136, 145, 147, 148, 178, 180; biochemical basis for, 127; reasons why it lacks precision and accuracy, 126

Luck: as a cause of environmental stability, 69

Marine chemistry: anoxic zones, 28;

depletion of surface nutrients by photosynthesis, 28; determined by steady-state considerations rather than thermodynamic equilibrium, 76; nutrient/depth profile, 29
Marine detritus: organic carbon adsorbed on mineral surfaces, 85; removal of nutrients from the photic zone, 28, 29, 84, 85, 89
Mars: compared to Earth and Venus, 2, 70, 109; Viking probe to, 109
Mauna Loa: records of atmospheric CO_2, 25, 39, 80, 81, 103, 104
Metabolism, cellular: "crystallized," 148; demands complementariness among genes, 134; genetic origins of regulation, 132, 148; ground plan, 15; orderliness of, 15, 103, 122, 127; regulation related to fitness, 133, 148
Methane: increase in the atmosphere, 49; sources and sinks, 50, 51, 61
Michaelis constant: = 1/association constant for binding of ligand by protein, 128; of growth for nutrient, 149, 176; of phosphate transport for PO_4, 90; of ribulose-1,5-bisphosphate carboxylase for CO_2, 92; of terminal oxidases for O_2, 84, 88
Milankowitch cycles, 11
Milieu intérieur. *See* Environment, internal

Negative feedback: as a cause of environmental stability, 69, 75, 112; control of the internal environment, 120; distinction between geochemical and biochemical, 113, 117, 130; in enzymic systems, 114; mechanisms may be vulnerable to anthropogenic pressure, 113
Net primary production (NPP). *See* Production
Nitrate reductase: active and inactive forms, 168; distinction between assimilatory and dissimilatory, 167; interaction with the gene product of *nit* 4/5, 169; mutations in the structural gene for, 169; production repressed by glutamine, 170; reducing equivalents for, 168, 171; regulated by CN?, 168; regulation of, 167; regulation of the gene for, 168; role in nitrogen cycle, 157
Nitric oxide: generated by bolide impact, 6; significance in atmospheric chemistry, 51
Nitrification, 31, 54, 59
Nitrite reductase, 167
Nitrogen: assimilation and photosynthesis, 153; fixation by cyanobacteria, 171; fixation by symbiotic systems, 171; fixation, enzymic reactions of, 172; fixation, role of sulfated oligosaccharides as signal molecules in rhizobial symbioses, 176; global fixation rate, 33, 53; global rates of incorporation into biomass, 32; industrial reduction to NH_3, 53; mobilization, 32; multiple valence states, 31; oxidation to NO_x, 53; pool sizes, 32, 156
Nitrogen cycle: biomolecular links to the sulfur cycle, 175; enzymic processes of, 154; four-box model, 52, 154
Nitrogenase, 33; dinitrogen reductase, regulated by ADP-ribosylation, 173; dinitrogenase reductase, structure of, 172; dinitrogenase, structure of, 172; energy requirements of, 172; flow of reducing equivalents through, 172; obligatory production of H_2 in the catalytic cycle of, 172; overall reaction, 172; regulated by O_2, 175; regulation of the genes for, 174; role in nitrogen cycle, 157; role of P_{II} in the genetic regulation of, 174; stoichiometry of ATP hydrolysis and electron transfer, 172
Nitrous oxide: as a source of stratospheric NO_x, 62; atmospheric half-life, 53, 155; sources and sinks, 54
NR_{II}. *See* P_{II}

Ocean: as sink for CO_2, 47, 97
Oxygen: atmospheric levels could be a fortuitous consequence of the ferrous minerals of the early Earth, 71; atmospheric turnover time, 83; consumed by respiration, 26; dependence of carbon burial rate on pO_2, 83; dependence of removal rate on pO_2, 83; factors affecting long-term

Index

stability, 86; factors determining the steady-state level, 87; measurement of seasonal fluctuations, 82; Michaelis constants of terminal oxidases for, 84, 88, 150; photosynthetic production, 21; possible variations during the Phanerozoic era, 8, 9, 67; production in stoichiometric equivalence to burial of organic carbon on the ocean floor, 9, 29, 67, 82–89; reducing sinks for, 73, 82; stability of the contemporary level, 73, 81; steady-state level determined by oceanic ventilation, 87; steady-state level determined by the marine biogeochemical cycle of phosphorus, 89–90

Ozone: Antarctic "hole," 62; as a filter of UV radiation, 62; causes of stratospheric depletion, 62; destroyed by bolide impact, 6; pathways of formation, 60; tropospheric sources, 60

Phosphate: Michaelis constant for transport of, 90, 127; rate of mining, 55; removal from the ocean, 30, 89, 90

Photorespiration, 23, 93, 95

Photosynthesis: at the Cretaceous-Tertiary boundary, 7, 100; biochemistry of, 20–21; discrimination against ^{13}C, 8, 45, 65; energy yield, 24; in C-3 and C-4 plants, 93, 95; limited by CO_2, 78, 92; marine, nutrient limitation, 29, 84; oxidized products, global inventory, 70; relationship to nitrogen assimilation, 153; stoichiometry of CO_2 assimilation and O_2 production, 24, 80

Physiological variation: and fitness, 130, 132; and genotype, 131; as a phenotypic characteristic, 130

P_{II}, regulatory protein: allosteric regulation of the uridylylation/deuridylylation of, 160; deuridylylated form activates NR_{II} phosphatase, 163; extent of uridylylation influenced by 2-oxoglutarate/glutamine ratio, 160; in the regulation of glutamine synthetase, 160, 161, 163; uridylylation leads to activation of the genes for glutamine synthetase, 164; uridylylation leads to activation of the genes for nitrogen fixation, 174

Production: estimation of global, 23, 24; gross primary, 22, 41; limitation by nutrients, 149; net primary, 22, 41; oceanic, limitation by nutrients, 48, 152; ratio of GPP to NPP, 23

Productivity, 23

Proteins, 3-dimensional structure, 122; active site is a feature of, 127; as phenotype, 128; determination by X-ray diffraction, 127; determined by 2-dimensional sequence of amino-acids, 128; environmental sensitivity, 128; evolutionary selection, 122, 129; mutational origins, 122; prediction from amino-acid sequence, 129; specificity of ligand binding, 118, 120, 122; stereochemical complementarity to ligand, 128

Proton: input to regional ecosystems, 59

Purposiveness: as a characteristic of living organisms, 135; evolutionary account of its origin, 135

Rate-limiting reactions: identification of, 116; in metabolic control, 116

Redfield ratios, 28, 56, 85, 152; reconciliation with Liebig's Law, 153, 166; variability of, 152

Reductionism, 180

Respiration, 25; dark, 19, 22, 26; heterotrophic, 19, 26; microbial, 27; phytoplankton, 28

Ribulose- 1,5-bisphosphate carboxylase (Rubisco), 20, 92

Self-organization, 144, 147; and the laws of complexity, 145; of geochemical systems, 106

Size: as a cause of environmental stability, 69, 72, 80, 81

Solar luminosity: at the origin of life, 76, 77

Stability: a condition for habitability, 2, 105, 112, 147; and concern about human impact on the global environment, 14, 72, 36, 37, 113; mechanisms responsible for, 69; multiple states, 14; of biological macromolecules, 127, 146

Steady-state systems: composition determined by kinetic characteristics, 79, 147; distinction from equilibrium systems, 80; stability of, 75, 106

Sulfate ^{34}S: changes negatively correlated to ^{13}C changes, 12
Sulfate reduction: in anaerobic sediments, 33; isotopic fractionation during, 7; stoichiometrical relation to carbon oxidation, 34
Sulfides: in marine sediments, 33, 34; isotopic shifts caused by oxidation of, 8
Sulfur: anthropogenic emissions counteracting the greenhouse effect, 100; anthropogenic sources, 57; gaseous forms, 33, 57; mineralization, 33; multiple valence states, 33; rate of mining, 57

Teleonomy, 135, 141, 143
Temperature: abiotic and biologically controlled global thermostats, 77, 92; effects on soil carbon, 96, 106; expected global change, 64, 99, 140; ineffectiveness of global thermostat compared to physiological regulation, 105; of the early Earth, 77; Pleistocene, 10; rapid changes during the last glacial-interglacial transition, 13, 98; stability on Daisyworlds, 140
Tree rings: carbon isotopes in, 44, 94
Turnover time: definition, 25; oceanic carbon, 96; phosphorus pools, 30, 85; relationship to pool size, 72
Two-component regulators, 165; in eucaryotes, 166

Uniformitarianism, 3
Uridylyl removing/uridylyltransferase. See P_{II}

Venus: compared to Earth and Mars, 2, 70, 109
Verhulst equation, 151

Water: isotopic composition, 10
Weathering: effects of biological activity, 91; effects of temperature and rainfall, 77, 91; oxygen consumption in the oxidation of reducing materials in rocks during, 82